RISK ANALYSIS AND SCIENTIFIC METHOD

D0813730

K. S. SHRADER-FRECHETTE

Dept. of Philosophy, University of Florida

RISK ANALYSIS AND SCIENTIFIC METHOD

Methodological and Ethical Problems with Evaluating Societal Hazards

D. REIDEL PUBLISHING COMPANY

A MEMBER OF THE KLUWER ACADEMIC PUBLISHERS GROUP

DORDRECHT / BOSTON / LANCASTER

Library of Congress Cataloging in Publication Data

Shrader-Frechette, K. S., 1944–
Risk analysis and scientific method.

Bibliography: p.
Includes indexes.
1. Technology assessment. 2. Decision-making.
3. Risk. I. Title.
T174.5.S48 1985 333.7′1 84–17985
ISBN 90–277–1836–9
ISBN 90–277–1844–X (pbk.)

Published by D. Reidel Publishing Company,
P.O. Box 17, 3300 AA Dordrecht, Holland.

Sold and distributed in the U.S.A. and Canada
by Kluwer Academic Publishers,
190 Old Derby Street, Hingham, MA 02043, U.S.A.

In all other countries, sold and distributed
by Kluwer Academic Publishers Group,
P.O. Box 322, 3300 AH Dordrecht, Holland.

All Rights Reserved
© 1985 by D. Reidel Publishing Company
No part of the material protected by this copyright notice may be reproduced or utilized in any form or by any means, electronic or mechanical, including photocopying, recording or by any information storage and retrieval system, without written permission from the copyright owner.

Printed in The Netherlands

For Eric

TABLE OF CONTENTS

PREFACE

Much of the work in this volume was supported by the National Science Foundation under Grant SES82-05112 from the Program in History and Philosophy of Science and the Division of Policy Research and Analysis. (Any opinions, findings, conclusions, or recommendations expressed in this publication are those of the author and do not necessarily reflect the views of the National Science Foundation.)

Several of these essays were written because of the impetus afforded by speaking invitations. An earlier version of Chapter 3 was presented in Berkeley in January 1983 at a Principal Investigators' Conference sponsored by the National Science Foundation, Division of Policy Research and Analysis, Technology Assessment and Risk Assessment Group. In May 1982, an earlier version of Chapter 5 was presented at the meeting of the Society for Philosophy and Technology, held in conjunction with the American Philosophical Association meeting, Western Division, in Columbus, Ohio. Finally, earlier versions of Chapter 6 were presented in Boston in December 1981 at the Boston Colloquium for the Philosophy of Science, as well as at the University of Delaware in January 1982 and at the Biennial Meeting of the Philosophy of Science Association held in Philadelphia in October 1982. An earlier version of this same chapter was published in *Philosophy of Science Association 82*, volume 1, ed. T. Nickles, Philosophy of Science Association, East Lansing, Michigan, 1982.

A number of people have helped to make this book better than it might have been. Betty Myers Shrader provided flawless proofreading and editorial assistance, and a number of colleagues in philosophy, mathematics, economics, and technology assessment have guided me with their insights. I am especially grateful to Joseph Agassi, Christopher Belshaw, Ed Byrne, Stan Carpenter, Bob Cohen, Paul Durbin, Ron Giere, Sheldon Krimsky, Ed Lawless, Doug MacLean, Joshua Menkes, Alex Michalos, Toby Page, Maurice Shrader-Frechette, and Stuart Spicker, each of whom has provided constructive criticisms of earlier versions of one or more of the chapters. Whatever errors remain are my responsibility.

My greatest debt is to my husband, Maurice, the brightest and most loving critic of my work, and to our children, Eric and Danielle. They make everything worthwhile.

The University of California, Santa Barbara K. S. S.-F.
June 1984

PART ONE

INTRODUCTION

CHAPTER ONE

OVERVIEW

1. THE RISK DILEMMA

In the fifteenth volume of his *Annals*, Tacitus recognized a dilemma. He noted that "the desire for safety lies over and against every great and noble enterprise." The task of risk analysis is to resolve this dilemma, to help us discover:

> how safe is safe enough;
> how much we ought to pay for safety;
> how equitably we ought to distribute societal risks; and
> how reliable are our scientific measures of risk.

Resolving the risk dilemma is difficult because every gain in safety has been won at the expense of time and money, and every gain in efficiency and technological progress has been won at the expense of human and environmental risk. The United States, in particular, has advanced in part because numerous persons in the past either chose, or were made, to forego personal concern for safety in order to achieve some broad, societal goal. Railroads, for example, linked one coast to another and set the stage for industrial achievement and burgeoning commerce. While this great enterprise was being accomplished, however, one writer noted that the "human machine was driven to the limit without lubrication or repair [then] ... simply scrapped when disease, often the direct result of the occupation, robbed it of further usefulness."[1] As one observer at the turn of the century put it: "war is safe compared to railroading in this country."[2]

Worker fatalities on U.S. railroads reached a peak in 1904

with a rate of 28 per 10,000 employees. Coal mining was even worse. In the early years of the twentieth century, while the average annual death toll in the mines of major European coal-producing countries varied from 9 to 13 fatalities per 10,000 workers, the United States average was between 33 and 34 deaths per 10,000 workers.[3]

Although in the past decade numerous laws have been passed and many federal agencies have been created to regulate health and safety, the threat of occupational and public risk still looms large. In occupational areas alone, at least one U.S. worker in 40 annually succumbs to death or to a reportable injury on the job. In one year, accidents at U.S. workplaces cause more than 14,000 deaths and more than 100,000 permanent disabilities. The U.S. Public Health Service estimates that approximately 400,000 new cases of occupational disease appear annually in the U.S., and up to 100,000 occupational disease-induced deaths occur each year.[4] These statistics, of course, do not even include the risks to the general public associated with various environmental and technological hazards.

2. PROBLEMS WITH RISK ANALYSIS

Recognizing the enormity of the hazards we face, decision theorists, philosophers, economists, and physical scientists began an intensive effort, approximately twenty years ago, to identify, estimate, and evaluate the various risks to which citizens are exposed. At least in the U.S., much of this effort was a consequence of new legislation, e.g., the 1969 National Environmental Policy Act, and a result of new regulatory agencies, e.g., the Occupational Safety and Health Administration (OSHA), which was created in 1970. Over the years, this new area of research has come to be known as risk assessment or risk analysis.

Because of the lack of statistical data and the dearth of explicit standards that risk assessments must meet, in order to be used as evidence in policy decisions, experts have had few guidelines for undertaking risk analyses.[5] Aware of the methodological problems associated with inadequate standards for the practice of this "developing science" of risk assessment,[6] committees of the U.S. National Academy of Sciences, the U.S. National Research Council, and the U.S. National Science Foundation attempted in the early 1980's to establish research priorities in risk analysis. The committees identified substantive problems in need of risk assessment, e.g., fluoridation and acid rain as well as issues of risk-assessment methodology in need of philosophical analysis, e.g., discounting, equity in shadow pricing.[7]

At the same time, the United States government recognized the need for reliable risk assessment and passed the Risk Analysis Research and Demonstration Act of 1982 (H.R. 6159). The purpose of this act was to establish a program under the coordination of the Office of Science and Technology Policy for improving the use of risk analysis by those federal agencies concerned with regulatory decisions related to the protection of human life, health, and the environment. The act noted that such regulatory decisions "involve scientific, economic, social, and philosophical considerations" and that "improving the quality of the scientific component of such decisions is an important aspect of preserving the public's freedom of choice in weighting the social and philosophical values intrinsic in them".[8] Hence the act stressed the necessity of government agencies' "securing information on, and an understanding of, the scientific bases for their regulatory decisions".[9] In order to secure this information, the Risk Analysis Research and Demonstration Act was intended to encourage researchers to "define criteria and standards to guide the development and use of risk analysis" and "to improve the methodologies" used in risk analysis.[10]

3. THE SCOPE OF THE PRESENT VOLUME

One way to improve risk-assessment methodologies is to subject them to philosophical analysis in order to clarify and evaluate their scientific, logical, epistemological, and ethical presuppositions. As is pointed out in the next chapter, the two most prominent methods of risk evaluation are risk–cost–benefit analysis and the method of revealed preferences. Both of these approaches are obvious targets for the analysis of risk-assessment methodologies which has been requested by numerous academic and governmental committees as well as by the legislators who passed the Risk Analysis Research and Demonstration Act of 1982. Since a significant amount of attention has already been devoted to investigation of the methodological presuppositions implicit in risk–cost–benefit analysis,[11] this volume is dedicated to philosophical analysis of some of the methodological presuppositions implicit in the revealed-preferences approach to risk evaluation. Although these analyses do not come close to exhausting the work that needs to be done to clarify and improve the scientific, logical, epistemological, and ethical underpinnings of the method of revealed preferences, they are a first step. Each subsequent chapter of this volume focuses on a philosophical problem which is central to the enterprise of risk assessment. If the arguments about the inadequacies of these methodological presuppositions are correct, then extensive changes will need to be made in the philosophical theories underlying risk analysis, the scientific techniques used to practice it, and the public policy decisions implemented as a consequence of it.

As a first step in investigating the scientific, logical, ethical, epistemological, and ethical foundations of the risk-assessment method known as revealed preferences, this volume has a number of limitations which ought to be acknowledged at the outset. *First*, and most obvious, it covers only a small number of the problems of scientific method associated with various risk-assessment frameworks and techniques. *Second*, because of the particular experience of the author with governmental, industrial, and university risk-assessment teams, the book is limited to United-States problems (e.g., the adequacy of the theory of the compensating wage differential) which usually do, but sometimes do not, have application to risk analysis as practiced in other countries. *Third*, since the focus of this work is on problems of scientific *method*, actual case studies dealing with alternative technological hazards are employed only to the extent that they raise particular methodological issues. Although this volume discusses particular risks (e.g., from nuclear technology, liquefied-natural-gas technology, and pharmaceuticals), such examples can be extended to many other technologies because the same methodological issues usually arise in considering the risks they pose. No exhaustive attempt has been made, however, to evaluate the hazards associated with the myriad sources of contemporary risk. *Fourth*, although the work discusses certain methodological problems associated with particular risk-assessment techniques, it should not be assumed that all the methodological presuppositions evaluated here are held by all members of the risk-analysis community. In the case of the linearity assumption (discussed in Chapter Six), for example, a number of prominent assessors, such as Cohen, Comar, Gibson, Lee, Maxey, Okrent, Rasmussen, Rudman, Starr, and Whipple, do subscribe to this presupposition. Not everyone does, however. Regardless of the universality of adherence to particular tenets, such as the linearity assumption, my strategy has been to evaluate methodological presuppositions which are widely held by prominent assessors and which have heretofore not been subjected to analytical, philosophical scrutiny.

3.1. *Risk Assessment*

The second chapter of this volume sets the stage for the methodological discussions of later chapters. Its purpose is to outline the steps in risk assessment or risk analysis so that one can understand the myriad ways in which faulty or misused scientific methods can jeopardize the correctness of assessment conclusions and therefore the appropriateness of risk-management policies. Although there are differences in the ways in which risk assessments are performed in various countries, they generally include three steps: risk identification, risk estimation, and risk evaluation.

Risk identification is accomplished by means of various scientific methods, especially those common in toxicology and epidemiology, and its results are dependent upon the use of a number of biostatistical techniques. Once a particular risk is identified, more sensitive toxicological and epidemiological methods are used to estimate the magnitude of the risk. This involves determining the dose–response relationship and estimating the dose received by particular populations. The third stage of risk assessment is risk evaluation, determining whether a given risk is acceptable. To determine risk acceptability, assessors generally employ one or more of the following approaches, the first two of which are the most widely used: risk–cost–benefit analysis; the method of revealed preferences; the method of expressed preferences; and the method of natural standards. Each of these methods will be explained and evaluated in the next chapter.

After outlining the various steps involved in the three stages of risk assessment (risk identification, estimation, and evaluation), Chapter Two surveys the main scientific, logical, epistemological, and ethical difficulties associated with the methods employed in each of the steps. Most of these problems arise as a consequence of assessors' employing questionable scientific theories or doubtful epistemological assumptions.

3.2. *Two Ethical Problems Associated with the Method of Revealed Preferences*

The survey of risk assessment techniques having been accomplished in Chapter Two, the stage is set for discussion of specific assumptions inherent in techniques of risk evaluation. Chapters Three and Four are devoted to analysis of two *ethical problems* associated with the method of revealed preferences, while Chapters Five and Six deal with two *decision-theoretic problems* associated with the method.

Chapter Three, 'The Commensurability Presupposition', is an analysis and evaluation of one of the most important equity-related assumptions routinely employed in risk evaluation. This is the assumption that the marginal cost of saving lives, across opportunities, ought to be the same. In Chapter Three, I investigate three claims often made to support this assumption, which I call the commensurability presupposition. These are the equity claim, the rationality claim, and the responsibility claim. Next I evaluate several criteria in terms of which the use of the commensurability presupposition may be judged acceptable or unacceptable. I argue that, while use of this presupposition is probably desirable in specific types of 'easy cases', it is unacceptable in so-called 'hard cases'. Hence, although the commensurability presupposition ought not to be accepted in all instances, I claim that two principles help to determine the cases in which its use is acceptable. These are the principle of *prima facie* egalitarianism and the principle of everyone's advantage.

In Chapter Four, 'Occupational Risk and the Theory of the Compensating Wage Differential', I discuss the widespread belief, especially among proponents of the method of expressed preferences, that a double standard for worker and public safety is ethically defensible on the grounds that workers consent to, and receive compensation for, the higher risks that they bear. After analyzing four arguments made on behalf of the theory of the compensating wage differential, viz., the welfare argument, the market-efficiency argument, the autonomy argument, and the exploitation-avoidance argument, I consider several situations in which there might be grounds for rejecting the double standard for occupational and public risk. One situation occurs whenever a worker's acceptance of an occupational risk involuntarily imposes additional risk on someone who has not consented to it. The other situation occurs whenever risk assessors adopt inconsistent stances toward risk perceptions in order to support the existence of a double standard. I close the chapter by discussing conditions for consent, compensation, and voluntary choice which, when met, might render the theory of the compensating wage differential ethically acceptable. I reaffirm the thesis that risk assessors' acceptance of the theory of the compensating wage differential, in its current form, errs by virtue of confusing market demands with ethical justifications.

3.3. *Two Decision-Theoretic Problems Associated with the Method of Revealed Preferences*

In addition to the problems of equity, informed consent, and compensation which face risk assessors who are concerned about the ethical presuppositions of their work, there are also a number of decision-theoretic difficulties associated with central scientific, logical, and epistemological presuppositions of risk-analysis methodologies. Chapters Five and Six focus on two problems in the latter set.

Chapter Five, an analysis of what I call the 'probability-threshold position', deals with the widespread belief, among practitioners of the method of revealed preferences, that a certain amount of risk is acceptable. The probability-threshold position is the view that one ought to ignore any small risks (those, for example, for which the individual probability of fatality is at the threshold of 10^{-6} per year or less). After a brief survey of two alternatives to the probability-threshold approach, I formulate and evaluate three arguments made on behalf of this position. These are the argument from decision theory, the argument from ontology, and the argument from epistemology. After analyzing the flaws in each of these arguments, I investigate a rejoinder for which Kenneth Arrow is famous. I show that, while Arrow indeed proves that some ordering theory justifies risk assessors' counting subthreshold probabilities as zero, it is impossible to relate this ordering to what is good, preferable or desirable in any real sense. Hence Arrow's rejoinder does not solve the problem of whether the probability-threshold position is part of ethically desirable social policy. I conclude by arguing that, since proponents of the probability-threshold position ignore the fact that various ethical parameters affect the acceptability of subthreshold risks, risk-assessment parameters ought to be weighted in terms of alternative ethical criteria. In this way, the normative consequences of assessment presuppositions could be rendered explicit and thus available for public debate.

In Chapter Five, 'The Linearity Assumption', I argue against a central presupposition of the method of revealed preferences. This is the assumption that there is a linear relationship between the actual probability of fatality and the value of avoiding a social risk or the cost of a social risk. The main object of this final chapter is to show that the methodological, logical, scientific, and epistemological underpinnings of the linearity assumption are highly questionable. As a consequence of my arguments on this point, I maintain that assessors ought to give more consideration to broadening their interpretations of 'unit cost' and 'societal risk' and to abandoning their claims about linearity. I return, finally, to a theme of the previous chapter and suggest that a system of ethical weights be applied to various assessment parameters so as to avoid the simplistic risk evaluations generated by the linearity assumption.

4. NEW DIRECTIONS

As this overview of the methodological issues to be discussed in this volume indicates, I believe that risk assessment needs to be improved in some significant ways. Most basically it needs to avoid simplistic or reductionistic approaches to problems of safety and to address the complex epistemological, logical, and ethical problems raised by questions of evaluating risk. In terms of the method of revealed preferences, risk assessors need to forego appeal to the commensurability presupposition, the compensating wage differential, the probability-threshold position, and the linearity assumption and, instead, to investigate the ethical and methodological constraints which, in a particular situation, determine whether these appeals are philosophically defensible.

In identifying some of the scientific, epistemological, and ethical limitations of key assumptions within the method of revealed preferences, my goal is not only to clarify the status of various claims made within the risk-assessment community but also to reveal the numerous and unrecognized ways in which values infect risk analysis. Philosophers of science have helped to banish the myth of complete objectivity from traditional sciences such as physics and biology. The groundwork for accomplishing the same task within risk assessment has barely begun. My hope is that in beginning this groundwork we will slowly come to a clearer vision of the monumental problems of scientific method, epistemology, ethics, and political philosophy posed by the new discipline of risk analysis.

NOTES

1 Carl Gersuny, *Work Hazards and Industrial Conflicts*, University Press of New England, London, 1981, p. 20; hereafter cited as *Hazards*.

2 Quoted by Gersuny, *Hazards*, p. 20.

3 Gersuny, *Hazards*, p. 20. See also E. A. Crouch and R. Wilson, *Risk/Benefit Analysis*, Ballinger, Cambridge, Massachusetts, 1982, pp. 12, 26, 40, 42, 44; hereafter cited as: *RBA*.

4 E. Eckholm, 'Unhealthy Jobs', *Environment* **19** (6), (August/September 1977), 29; hereafter cited as: Jobs.

5 H. Kunreuther and E. Ley, 'Overview', in *The Risk Analysis Controversy* (edited by Kunreuther and Ley), Springer-Verlag, New York, 1982, p. 4, make this same point; hereafter cited as: *RAC*.

6 S. Levine, 'Panel: Use of Risk Assessment', in *Symposium/Workshop . . . Risk Assessment and Governmental Decision Making* (edited by the Mitre Corporation, McLean, Virginia, 1979, p. 634, is one of those who believes that risk assessment is a science.

7 H. Raiffa, 'Science and Policy', in *RAC* (edited by Kunreuther and Ley), pp. 27–37.

8 *Risk Analysis Research and Demonstration Act of 1982*, Sec. 2. (a), 11–12, 19–22; hereafter cited as: RARDA.

9 RARDA, Sec. 2. (b), 3–4.

10 RARDA, Sec. 2. (b), 9–12.

11 See, for example, *Philosophy and Economic Theory* (edited by Frank Hahn and Martin Hollis), Oxford University Press, New York, 1979; John Harsanyi, *Essays on Ethics, Social Behavior, and Scientific Explanation*, Reidel, Boston, 1976; and K. S. Shrader-Frechette, *Science Policy, Ethics, and Economic Methodology*, Reidel, Boston, 1984, esp. Chapters 4–7.

CHAPTER TWO

RISK ASSESSMENT

1. INTRODUCTION

Fires, floods, droughts, famines, and infectious diseases, all natural hazards, were once the principal dangers faced by society. Today they have been replaced by human-created hazards such as nuclear war, toxic chemicals, liquefied-natural-gas explosions, and automobile accidents. Although our ability to predict and control our environment has increased immensely and rendered us much more able to prevent natural hazards, we are only beginning to learn whether, when, and how to control technological hazards.

Two of the more recent examples of the failure of technological risk management are Love Canal and Three Mile Island. Known victims of the toxic chemicals dumped at Love Canal by the Hooker Chemical Company are 949 families who have had to be moved from their homes near the site. Claims against Hooker now total approximately $11 billion, and it is not clear that all damage is known. Still uncertain is the effect of the more than 800 chemicals on the descendents of the Love Canal families, since many of the toxins are mutagenic as well as carcinogenic.[1] Unlike the Love-Canal catastrophe, the Three-Mile-Island Nuclear accident did not cause extraordinary health problems. However, the physical damage to the plant was devastating. It will cost at least $2 million to bring the reactor back into operation after four years of radiation clean-up, and some experts say that the minimum cost will be not $2 million, but $2 billion, to bring the reactor back on line. This is several times more than the original cost of construction.[2]

2. REASONS FOR THE FAILURE TO MANAGE TECHNOLOGICAL RISKS

Many other failures in managing technological risks could be discussed – nerve gas stored on the Denver flight path, chemicals in Christmas cranberries, carcinogens in children's bedclothing, and the problems with the Fermi Breeder Reactor that allegedly almost destroyed the city of Detroit. Although these failures must be seen in the perspective of thousands of technological successes wrought in the areas of medicine, energy, transportation, and communication, and although technological advances have likely given us a far better life than our ancestors enjoyed, the question remains. Why have we not done better in managing our technological risks?

One reason for our failure is that *conflict of interest* has pervaded the regulatory process. Safety concerns at the U.S. Nuclear Regulatory Commission, for example, often have been overwhelmed by the need to protect and promote nuclear technology. The government coverup at the early stages of the Three-Mile-Island accident made this abundantly clear.[3] Likewise, at Love Canal, to take another example, conflict of interest inhibited society's ability to manage the risk. On the one hand, Hooker Chemical Company denied that their Love-Canal dump presented a health hazard, even long after company scientists affirmed this fact; on the other hand, environmentalist zealots who needed accurate epidemiological data were too eager to publicize unsubstantiated and uninterpretable findings.[4]

Public anxiety is another reason why we have often failed to manage our technological risks in a rational way. Sometimes the result of uncertain science and misunderstanding, public fear often blocks effective implementation and management of important technologies. Appropriate risk management requires that government be neither paralyzed by groundless public anxiety nor unconcerned with legitimate fears. Risk management is hindered whenever the one is confused with the other.

A *third reason* for our failure to manage our technological risks, and the reason with which I am most concerned here, stems from improper or uncertain scientific methodology and from incorrect use of scientific methodology in assessments of technological risks. Any risk assessment is only as good as the methodology underlying it. This means that if the methodology is flawed, then so is the assessment. And if the assessment is flawed, then so is the public policy made on the basis of it. For example, many reputable groups of scientists, including the American Physical Society, have criticized the scientific and mathematical methodology of the most famous and complete U.S. risk assessment of commercial nuclear reactors. They charged that some of the allegedly low-risk probabilities and consequence (fatality) magnitudes were in error (because of faulty methodology) by as much as two and three orders of magnitude.[5] To the extent that this risk assessment, known as the Rasmussen Report, is in error, then to that same degree is public policy regarding the acceptability of nuclear risk also in error, since U.S. energy policy is in large part dependent on the results of this assessment.

3. STEPS IN RISK ASSESSMENT

To understand the myriad ways in which faulty or misused scientific methods can jeopardize the correctness of assessment conclusions and therefore the appropriateness of risk-management policies, it would be instructive to know how assessments are performed. This would provide some idea of the precise ways and the specific points at which uncertainty is likely to arise in the risk-assessment process.

International comparisons indicate that there are differences in the way risk assessments are carried out in various countries.

For example, approximately four times as many drugs have been approved for physicians' use in the United Kingdom as in the U.S., over the last decade.[6] This difference is explicable in part on the basis of variations in risk assessment strategies in the two countries. The U.S. tends to emphasize animal testing prior to therapeutic use, for example, while Great Britian usually employs only limited animal tests but emphasizes closely monitored therapeutic uses. Apart from minor discrepancies such as these, risk assessment as generally practiced throughout the globe tends to include three main processes: risk identification, risk estimation, and risk evaluation.

3.1. *Risk Identification*

For decisionmaking, the amount or severity of risk perceived is used as an approximation of the risk itself. 'Risk' is generally defined as a compound measure of the perceived probability and magnitude of adverse effect.[7] For example, one might say that, in a given year, each American runs a *risk*, on the average, of about one in 4,000 of dying in an automobile accident. Assessors most often express their measures of risk in terms of annual probability of fatality for an individual.

3.1.1. *Types of Risks Considered*

Although the first step in the three-part risk assessment process is risk identification, not all individual and societal risks which have been identified need be estimated and evaluated through some analytical framework. In practice, government, industry, citizens, and risk assessors are interested primarily in risks which are neither very large nor very small because these are the ones which involve societal controversy. Nearly everyone is already convinced that large risks ought to be avoided and that small ones are not worth worrying about. Moreover, it is usually not too difficult to place many events in one of three categories, on the basis of whether the risks they carry are very large, moderate, or small. This is because *historical risks* have had adverse consequences associated with them which have occurred often enough for data sufficient for analysis to have been accumulated. Historical risks include those from diseases, automobiles, industrial accidents, some forms of pollution, hurricanes, tornadoes, and lightning. *New risks*, however, include those arising from events never previously observed or those historical risks whose frequency is so low that it is hard to assess accurately whether they belong to the class of very large, moderate, or small risks. New risks include those such as reactor meltdowns and adverse consequences from exposure to previously unknown chemicals. Although assessors are interested in historical risks, and particularly in alternative models for drawing conclusions about the magnitude of historical risks, their primary focus is on new risks and on estimating and evaluating events likely to result in a moderate risk (events for which the annual probability of fatality, per person, is between 10^{-6} and 10^{-4}, for example).[8]

Inasmuch as studies of risk are often aimed at providing a basis for government regulation and risk policy, assessment is also directed primarily at investigation of societal, rather than individual, risks. *Individual risks* are those accepted through voluntary activities, e.g., smoking. These risks are assessed in terms of the individual's own value system, and each person has a relatively large degree of freedom in deciding whether to accept them. (Admittedly, however, government regulations, e.g., concerning smoking, limit the degree to which one's decision to take a risk is freely chosen.) *Societal risks* are generally involuntarily imposed rather than voluntarily chosen, although citizens often have some voice in the government or industry decision to impose them, e.g., siting a liquefied natural gas facility in a large population area. Unlike individual risks, societal risks are not assessed on the basis of each person's value system.[9] Rather, control over societal risks is generally in the hands of some government or political group. This means that accurate risk assessments are essential to reasoned societal decisionmaking, since every person's opinion on a given hazard cannot be followed, and since every citizen obviously cannot vote on every issue involving some sort of risk.

3.1.2. *Methods of Hazard Identification*

There are, for example, roughly 60 thousand commonly used chemicals, and approximately 1,000 new ones are introduced each year. In producing millions of materials and services, the U.S. industrial economy alone includes a labor force of more than 100 million workers, many of whom are exposed to a vast range of accidents and to numerous substances that could lead to acute or chronic disorders and to carcinogenic, mutagenic, and teratogenic disease. It is a substantial problem simply to identify the possible health hazards facing U.S. workers, let alone the general public.

Five methods are commonly used to identify hazards. These are: (1) use of case clusters; (2) comparison of compounds in terms of structural toxicology; (3) mutagenicity assays of simple test systems such as bacteria or cultured mammalian cells; (4) long-term animal bioassays; and (5) use of sophisticated biostatistical techniques for epidemiological analysis. These methods differ in terms of the information they yield, their length, and their cost.

3.1.2.1. *Use of case clusters.* Examination of case clusters is perhaps the oldest and most widely used method of identifying a hazard, and it has the lowest level of analytic sophistication. This method consists simply of noticing a number of cases of a rare disease, or an unusual concentration of cases of a common disease, and attempting to find the cause. Intuition is used to infer the possible cause and to examine relevant possibilities. For example, a century ago, Percival Pott inferred the cause of scrotal cancer among chimney sweeps. More recently, a number of physicians inferred the cause of liver cancer among vinyl-chloride workers. These examples reveal the power of identifying hazards on the basis of case clusters when the disease observed is otherwise relatively rare.[10]

This method is much less powerful when the health condition observed is more common among the population. Coke-oven gases, for example, were not identified as hazards earlier because exposure to them caused no unique disease and because the incidence of lung cancer which they caused was not significantly above that typically observed in the general population.[11] Another obvious difficulty with the method of case clusters is that the population at risk is often unknown; this means that the risk could arise from an occupational exposure, an environmental exposure, or some complex set of personal habits and characteristics, e.g., being a smoker, being under age 30, and being a user of oral contraceptives. Because the population at risk is rarely known in great detail, and because the method cannot control confounding factors, it never yields conclusive evidence. Rather, use of case clusters provides a way of obtaining 'hunches' to be checked out by more analytic procedures.

3.1.2.2. *Comparison of compounds in terms of structural toxicology*. A second method of hazard identification consists of comparing an agent's chemical or physical properties with those of known carcinogens in order to obtain some evidence of potential carcinogenicity. For example, many of the coal tars are known to be carcinogenic for humans. For this reason, one might use the method of structural toxicology to determine whether some untested coal tar, because of its similar chemical structure, is likely also to be a human carcinogen. Indeed, experimental data support such associations for a few structural classes. The main deficiency of this method, however, is that comparisons of compounds on the basis of structural toxicology are used best to identify potential carcinogens which ought to be the subject of future research. Hence this method is better used for priority testing for carcinogenicity testing than for actual determintion of carcinogens.[12]

3.1.2.3. *Mutagenicity assays of simple test systems.* A third method of identification is designed to reveal a possible carcinogenic hazard on the basis of a positive response in a mutagenicity assay. A great body of experimental evidence supports the belief that most chemical carcinogens are mutagens, and that many mutagens are carcinogens. In order to obtain data on possible carcinogenicity, assessors often employ short-term, *in vitro* testing, using either simple systems such as bacteria or cultured mammalian cells, in order to identify mutagenic effects as well as cell transformations. These laboratory tests are quick and relatively inexpensive, and they can be used to screen thousands of chemicals.

The main deficiency of the short-term mutagenicity assays is that the data they generate are rarely, if ever, sufficient to support a conclusion that an agent is carcinogenic. Hence these tests are valuable only for identifying potential carcinogens and for lending support to observations from animal and epidemiological studies. Moreover, since these tests have shown that more than 200 chemicals are mutagenic, and since all 200 cannot possibly be subjected to epidemiological survey over the short term, the mutagenicity assays can be used only in a limited sense to further the regulatory process.[13]

3.1.2.4. *Long-term animal bioassays.* The most commonly available data in hazard identification are obtained from long-term animal bioassays. These laboratory experiments are time-consuming and expensive, and they are usually performed on rodents. Their purpose is not to explore possible associations between agents and disease, but to test hypotheses, often about carcinogenicity. Consistently positive test results in the two sexes and in several animal strains and species, as well as higher incidences at higher doses, constitute the best evidence that a given substance is a carcinogen. In general, animal bioassays have proved to be reliable indicators of disease, and they will probably continue to play an important role in efforts to identify carcinogens.

There are, of course, a number of methodological problems associated with using long-term animal bioassays. The most obvious difficulty is with the inference that results from animal experiments are applicable to humans. Although this inference is fundamental to toxicological research, and although most cancer researchers accept it, there are occasions on which animal observations are not of obvious relevance to conclusions about human carcinogenicity. A more practical limitation of this method is that often, because of the nature of many carcinogenic effects and the limits of detection in animal tests, experimental data leading to a positive test result frequently barely exceed a given statistical threshold. Hence interpretation of the animal data may be difficult.[14]

3.1.2.5. *Use of sophisticated biostatistical techniques for epidemiological analysis.* A fifth class of methods for hazard identification comes from epidemiology, a more sophisticated form of case-cluster analysis. The goal of these biostatistical epidemiological studies is to show a positive association between an agent and a disease. This association is generally accepted as the most convincing evidence about human risk. The evidence tends to be convincing, in large part, because epidemiological analyses generally control for confounding factors in the experimental design.

Utilizing sophisticated biostatistical techniques, epidemiological studies may have either a descriptive or analytical orientation. That is, they may focus either on the distribution of a disease in a defined population or on the various factors associated with its incidence.

Two main approaches are used in epidemiological investigations, *retrospective* and *prospective*. Retrospective studies involve case studies of a group of persons who have a given disease and of a control group whose members do not have the disease. Epidemiologists obtain risk-related information for that disease by comparing the two groups on the basis of age, sex, genetic composition, occupation, place of residence, lifestyle, etc. Prospective studies involve following the medical histories of two different groups. One group is exposed to the potential disease-related substance, and the other (the control group) is not. Epidemiologists then compare the histories of the groups over a period of years.[15]

The main deficiency of epidemiological studies is that it is often difficult to accumulate the relevant evidence. This frequently occurs when a given risk is low, or when the number of persons exposed is small, or when the latency period (between exposure to the substance or agent and the onset of disease) is long, or when the exposures are mixed and multiple. This means that most epidemiological data require very careful interpretation. Apart from these problems, a remaining difficulty is that most chemicals in the environment have not been, and are not likely to be, tested using epidemiological methods (see Section 3.1.2.3 earlier).

3.2. *Risk Estimation*

Once a substance has been identified positively as a serious hazard, methods of epidemiology and toxicology can be used to estimate the magnitude of the risk. Risk estimation, the second step of risk assessment, involves two tasks. The first is to determine the dose–response relationship, and the second is to estimate the population at risk and the dose it receives from a particular substance.

3.2.1. *Determining the Dose-Response Relationship*

Determining the dose-response relationship, 'dose–response assessment' is the name given to the process of characterizing the relation between the dose of an agent administered or received and the incidence of an adverse health effect in exposed populations. The purpose of this method is to estimate the incidence of the effect as a function of human exposure to the agent. Dose–response assessment takes account of the intensity of exposure, the age pattern of the exposure, and perhaps other variables that might affect response, such as sex and lifestyle. Dose–response assessment usually requires extrapolation from high to low doses and from animals to humans. Hence, because of the inferences involved, all assessors should describe and justify the methods of extrapolation used to predict incidence. They should also characterize the statistical and biological uncertainties in these methods.[16]

In a very few number of cases, epidemiological data permit a dose-response relationship to be developed directly from observations of exposure and resultant health effects in humans. Even if these data are available, it is still usually necessary to extrapolate from the exposures observed in the study to lower exposures experienced by the general population. Since useful human data are absent for most chemicals being assessed for carcinogenic effect, however, dose–response assessment usually entails evaluating tests performed on rats or mice. In extrapolating from animals to humans, the doses used in bioassays must be adjusted to allow for differences in size and metabolic rate. Methods currently used for this adjustment carry the assumption that animal and human risks are equivalent when doses are measured as milligrams per kilogram per day; as milligrams per square meter of body-surface area; as parts per million in air, diet, or water; or as milligrams per kilogram per lifetime.[17]

In reality, there are a number of problems associated with making assumptions about the equivalence of human and animal risk. For one thing, metabolic differences can have important effects on the validity of extrapolating from animals to humans if, for example, the actual carcinogen is a metabolite of the administered chemical, and the animals tested differ from humans in their production of that metabolite. Other problems with drawing conclusions about human risks on the basis of animal bioassays are that species react differently, and that the absence of other environmental challenges excludes all interactions among challenges and thus may decrease resistance to the test substance and preclude assessors from making quantitative estimates for humans.[18]

Once assessors obtain fairly reliable animal data, they usually extrapolate by fitting a mathematical model to the animal dose–response information and then use the model to predict risks at lower doses corresponding to those experienced by humans. Currently, the true shape of the dose–response curve at several orders of magnitude below the observation range cannot be determined experimentally. The largest study on record, for example, can at best measure the dose corresponding to a one percent increase in tumor incidence, even though regulatory agencies are often concerned with much smaller increases in risk.[19]

A related problem with low-dose extrapolation is that a number of the extrapolation methods fit the data from animal experiments reasonably well, and it is impossible to distinguish their validity on the basis of goodness of fit. This means that low-dose extrapolation must be more than a curve-fitting exercise, and that considerations of biological plausibility must be taken into account. Plausibility, however, is hardly a clear criterion for assessing epidemiological models. Some scientists have challenged the practice of testing chemicals at high doses. They argue that animal metabolism of chemicals differs at high and low doses. For example, high doses may overwhelm a rodent's normal detoxification mechanisms and thus provide results that would not occur at the lower doses to which humans are exposed. Moreover, the actual dose of a carcinogen, for example, reaching the affected tissue or organ is usually unknown; this means that dose–response data is always based on administered, not tissue, dose.[20]

3.2.2. *Estimating the Population and Dose*

After an estimated dose–response relationship is arrived at for a substance, the next task of the assessor, at the risk estimation stage, is to determine the populations at risk and the dose they are likely to receive from the given substance. The first efforts at this second stage of risk estimation are to determine the *concentration* of the chemical to which humans are likely exposed. This may be known from direct measurement; usually, however, data are incomplete and must be estimated. This means that models must be used; they are usually complex, even when a particularly structured activity, e.g., workplace exposure, is being examined.

In the case of dose estimation for a workplace setting, assessment usually focuses on long-term airborne exposures. In the public or community environment, the ambient concentrations of a particular substance, e.g., an industrial chemical, to which people may be exposed can be estimated from emission rates from particular sources, but only if the transport and conversion processes are known. Various pollution-control mechanisms, however, require different estimates of the reduction in exposure that may be achieved.

Once information is developed on the pathways by which toxic substances reach people, then their concentrations in each pathway and the dose received by humans can be estimated or measured. Seemingly unimportant pathways can assume great significance, of course, because of food-chain and synergistic effects and because it is rare that a substance is uniformly distributed across pathways or across time. Given certain exposure pathways, the population at risk can be inferred from knowing, for example, who lives, eats, works, or breathes air at a particular site.

Even apart from normal differences in a substance's distribution across a pathway, there are other factors that create problems for accurate population and dose estimates. One of these factors is use. In the case of assessing chemicals present in food, for example, the use problem is particularly significant. Even when the amount of an agent in a food can be measured, differences in food-storage practices, food preparation, and dietary frequency often lead to wide variation in the type and the amount of the agent that particular individuals ingest. Even in nonfood cases, patterns of use affect exposure to numerous substances. For instance, a solvent whose vapor is toxic could be used outdoors or in a small unventilated room where the concentration of the toxin in the air is likely to be much higher. Another problem with estimation is that there may be a number of population groups whose members are especially sensitive to health effects of a particular substance. Pregnant women, those who suffer from allergies, children, or those leading sedentary lives may be affected much more adversely than are the average members of a population. Finally, even if the population at risk can be accurately estimated, a remaining problem is how to know what constitutes an adverse health effect from a particular substance. Obviously fatalities are adverse health effects, but what about minute, perhaps unperceived (but slightly measurable) physiological changes that, in themselves, are not disease? What counts as a precursor of disease? Numerous substances cause imperceptible changes, e.g., increases in airway resistance. It is unclear, however, even though such changes place a small strain on the body, whether they ought to be considered adverse and how important they are.[21]

3.3. *Risk Evaluation*

After a particular hazard has been identified and its risk estimated in terms of dose concentration and population exposure, the next (and final) stage of risk assessment is to evaluate the risk at hand. This requires determining whether a given risk is acceptable or ought to be judged acceptable by society. At this third stage of risk assessment, one must deal not only with scientific uncertainities, as in the two earlier stages, but also with normative controversies. It is at this last stage that the difficult problem of analyzing policy arises.

Risk assessors typically employ one or more of four methods of risk evaluation: (1) risk–cost–benefit analysis; (2) revealed preferences; (3) expressed preferences; and (4) natural standards. The first of these methods is formal, and well known to practitioners of welfare economics. The three remaining are informal methods.

3.3.1. *Risk–Cost–Benefit Analysis*

Formal methods of analysis attempt to clarify the issues surrounding evaluating the acceptability of risks through the application of well-defined principles of rationality. Risk–cost–benefit analysis and decision analysis are the most prominent formal modes of evaluating acceptable risk. Both methods proceed according to four main steps.

1. The risk problem is defined by listing alternative courses of action and the set of all possible consequences associated with each action.

2. The assessor next describes the relationships among these alternatives and their consequences. Various mathematical, economic, and social models may be used in the descriptions in order to arrive at quantitative accounts of dose–response relationships, market behavior, and event probabilities.
3. All the consequences of alternative risk decisions are evaluated in terms of a common unit. In risk–cost–benefit analysis (RCBA) this unit is money, and in decision analysis, the unit is utility, a measure of the probability of a consequence and the value attached to it.
4. All the components of the analysis are next integrated in order to produce a single number which represents the value of each alternative. In RCBA, this number represents the difference between the benefits of the decision alternative, on the one hand, and its risks and costs, on the other hand. In decision analysis, this final number represents the option's expected utility. The most desirable risk option is presumably that which (in the case of RCBA) has the greatest benefit to risk/cost ratio or that which (in the case of decision analysis) has the highest utility.[22]

If RCBA or decision analysis is interpreted as a *method* which is alone sufficient for determining acceptable risk decisions, then the risk option with either the greatest benefit to risk/cost ratio or the highest utility should be adopted. However, if RCBA or decision analysis is interpreted merely as one *aid* (among many) to decision-making about acceptable risks, then obviously it does not follow that anyone who uses these aids ought to subscribe to the alternative recommended by them.

Although RCBA goes by many different names, including

benefit–cost analysis, the label is used to refer to the explicit consideration of monetary advantages and disadvantages of one or more decision options. Currently RCBA is used by all U.S. regulatory agencies, with the exception of the U.S. Occupational Safety and Health Administration, OSHA, for routine decision-making. In fact, use of RCBA is required for all federal projects by virtue of a mandate of the 1969 U.S. National Environmental Policy Act.[23]

The most obvious deficiency in RCBA is that simply adding risks, costs, and benefits ignores who gets what. According to the criterion of the potential Pareto Improvement, or Kaldor–Hicks criterion (the principle underlying RCBA), an action is desirable provided that its benefits outweigh its risks and costs to a degree sufficient to allow the gainers to compensate the losers. Since, on this criterion, no compensation ever takes place, RCBA legitimates choosing the alternative that maximizes the difference between total benefits and total risks/costs, regardless of their distribution. In fact, one of the major criticisms of RCBA is that it fails to take account of distributive equity. Hence, to the extent that risks are evaluated in terms of RCBA, then to that same degree will they also fail to take account of distributive equity.[24]

Another problem with RCBA is that all risks, costs, and benefits cannot easily be translated into monetary units. Many philosophers, economists, and decision theorists claim that to attempt to translate everything into monetary units is to be guilty of 'economic philistinism'. A related problem is that expressing all risks, costs, and benefits in these monetary terms ties one to the existing system of market prices, with all the distortions arising from market imperfections, subsidies, failures to price nonmarket goods, monopolies, etc. This means that a risk assessment based on RCBA contains all the same discrepancies and imperfections as does the existing set of market distributions.[25]

By virtue of the fact that decision analysis requires that one measure the utility of a decision option in terms of probabilities and the value attached to them, decision analysis also exhibits a number of limitations whenever it is used for evaluating risk options. The most obvious deficiency is that there are numerous uncertainties about the present and future states of the world; hence it is difficult to assign a probability to various decision-theoretic options. Moreover, a more basic problem with the assignment of decision-theoretic probabilities is that decision analysts view probabilities as expressions of individuals' beliefs, not as characteristics of things. As a result, probabilities are elicited as judgments from the decisionmaker or expert. From a scientific point of view, there are numerous problems associated with relying on educated guessing and judgments, as opposed to calculation. Decision-theoretic probabilities used in risk evaluation could easily reflect the nonscientific prejudices of whatever experts make the probability judgment, e.g., in the case of the probability of a catastrophic nuclear reactor accident.

A related problem with decision analysis is that, unlike RCBA, its proponents do not quantify preferences by analysis of market data. Instead, decision theorists use subjective value judgments, or utilities, to measure the value of a particular risk outcome. By using subjective judgments, they are able to account for many factors not accommodated by the market, such as aesthetic preferences and risk aversion. The obvious problem with these subjective judgments, however, is that they can easily be criticized as arbitrary. For this reason, whenever more than one set of utility or probability judgments must be considered, decision theorists often prepare several complete analyses, each of which reflects the perspective of one party. Keeney and Raiffa (1976) recommend using a Supra Decision Maker when the various parties cannot agree.

Formal methods of risk evaluation, like RCBA and decision analysis, hold out the promise that the facts of a matter can be organized effectively and explicitly. Either type of analysis can, in principle, accommodate any fact or estimate, so long as it is compatible with the original problem definition. Moreover, by means of sensitivity analysis, once the RCBA or decision analysis is completed, theorists can look for places where a reasonable change in the structure, a utility, a probability, or a particular risk or benefit value, could lead to the selection of a different alternative. In this way RCBA or decision analysis can be corrected so as to provide more plausible results.

Perhaps the most basic objection to RCBA and decision analysis is that no formal methods are able to capture the nuances of risk evaluation situations. Proponents of this position maintain that there are no clearly specifiable criteria for determining an acceptable risk, and that typical risk assessment problems often faced by individuals involve no calculation of probabilities and consequences. Rather, they claim that risk decisions are made on the basis of intuition and 'know how', much as one learns to drive a car. Hence, they maintain that only use of intuition, subjectivity, or 'muddling through' intelligently will enable society to make decisions about acceptable risk.[26] For proponents of this object, any attempt to use a formal method of risk assessment merely obscures the difficult evaluative components of a situation. Moreover, since both RCBA and decision analysis have obvious deficiencies, they argue that neither is sufficient (whether alone or together) for evaluating risks, and that, at best, both are merely able to provide information which might did policymaking.

To argue, however, that RCBA and other formal methods provide no important basis for societal decisionmaking, and that intuition, democratic dialogue, and other nonstructured forms of policy analysis ought to be employed to evaluate risks, is to miss several important points. *First*, not to attempt to use some formal method of risk evaluation is to beg the question of whether anything can be learned from it. *Second*, although intuition and other subjective approaches may be adequate and perhaps even necessary for individual decisionmaking, opponents of formal methods forget that societal decisionmaking is far more complex. It requires some procedure (such as RCBA) in order to take account of diverse points of view, allow discussion among proponents of different approaches, and provide some clear and well-established basis for argument and agreement. Only a formal method has all these advantages. *Third*, simply because they do provide a clear framework for decisionmaking, methods of formal analysis lend themselves more easily to understanding by citizens and hence to democratic control. Whenever a method or process is not clearly spelled out, those who lack political power in the existing system are often 'cut out' of the decisionmaking process, simply because they do not know what is going on.[27]

3.3.2. *Revealed Preferences*

If one is wary of ambitious formal methods like RCBA or decision analysis, then an alternative might be to use an informal method. Whereas formal methods of risk evaluation are predicated on the assumption that we can rationally arrive at decisions about acceptable risk, informal methods are built on the presupposition that risks cannot be analyzed adequately in any short period of time. Rather, proponents of informal approaches believe that, although no *explicit* criteria are usually employed, society achieves an acceptable risk–cost–benefit trade-off over time, through a process of trial and error.

The most prominent of all informal approaches to risk evaluation is known as the method of revealed preferences. Practitioners of this method use the level of risk that has been tolerated in the past as a basis for evaluating the acceptability of present risks. Although they develop explicit calculations and specific decision rules in order to interpret historical risk levels, proponents of the method of revealed preferences do not believe that risk policies which have evolved without the benefit of careful quantitative analyses are incorrect or undesirable. For advocates of this method, historical policies may have prescriptive weight, even though they may neither have evolved, nor be justifiable, according to some formal, rational, decision rule. The method of revealed preferences is thus based on great faith in society's adaptive processes.

As a means of obtaining insight into risk evaluation, practitioners of the method of revealed preferences review historical patterns of the frequency of different consequences arising from a variety of causes. Most often, those who pursue this approach compare (1) the average annual probability of fatality associated with different activities; (2) the probability of fatality, per person-hour of exposure, for all participants in a given activity; or (3) the actual number of deaths per year associated with a number of activities. Chauncey Starr, for example, one of the founders of the method of revealed preferences, has compared the annual fatalities associated with events such as earthquakes, fires, floods, hurricanes, and tornadoes, and the deaths, per person-hour of exposure, associated with activities such as travelling by motor vehicle, private airplane, and commercial airline.[28]

Typically, Starr, Whipple, Cohen, Lee, Otway, Rudman, and others who employ the method of revealed preferences develop extensive tables of risks so that one can immediately determine the risk associated with anything from smoking 1.4 cigarettes to eating 100 charcoal-broiled steaks.[29] The purpose of such tables, as Cohen and Lee put it, is to insure that 'society's order of priorities' for risk reduction follows the ordering in the tables, from activities which are most risky to those which are the least so.[30]

Not all proponents of the method of revealed preferences appear to believe, as do Cohen and Lee, that the probability of fatality associated with a particular event is a sufficient basis for determining society's attitude toward reducing its risk. As discussed by Starr, this method improves upon simple comparisons of risk probabilities in that he considers the role of *benefits* in determining a number of decision rules for evaluating risks. Using the method of revealed preferences, Starr examined the relationship between the risk of death and the economic benefit associated with a number of events, technologies, and activities (where economic benefit is measured either in terms of money spent by the participant in the activity or in terms of the average contribution that the activity makes to one's income). On the basis of his calculated risk–benefit relationships, Starr formulated three hypotheses about the nature of acceptable risk. These propositions have since come to be known as 'laws of acceptable risk':

> The public is willing to accept voluntary risks roughly 1,000 times greater than involuntarily imposed risks.
>
> The statistical risk of death from disease appears to be a psychological yardstick for establishing the level of acceptability of other risks.
>
> The acceptability of risk appears to be crudely proportional to the third power of the benefits (real or imagined).[31]

Otway and Cohen performed a regression analysis on the same data base used by Starr and arrived at entirely different results. They claimed that, for *voluntary* risks in society, an allegedly acceptable risk is proportional to the 1.8 power of its benefits, and that for *involuntary* risks, an allegedly acceptable risk is proportional to the sixth power of benefits.[32] Starr, Otway, and Cohen nevertheless agreed that voluntary risks are more acceptable than involuntary ones and that, all things being equal, the greater the benefits involved, the more acceptable a particular risk is.

Apart from disagreements as to the precise nature of the risk–benefit relationship for various activities, the method of revealed preferences is controversial in more basic ways. Regardless of whether Starr or Otway and Cohen are correct about societal levels of risk acceptability, their use of past behavior to infer propositions about desirable risks raises a number of questions. Most of these queries have to do with the plausibility of various assumptions central to the method of revealed preferences.

Perhaps the most basic assumption in the method of revealed references is that past societal risk levels for various activities reveal correct or desirable policy decisions. Obviously, however, past risks may have been at a given level, not because society judged that level acceptable, but because greater safety was not obtainable at the time, or because there was inadequate knowledge of the risk. Or, perhaps given risk levels existed because regulation was too lax or too strict, or because individuals could not afford to expend the monies necessary for greater safety, or because the control technology for reducing risk was not available. In other words, many factors, such as income and social structure, may have determined risk levels in the past. Hence, it cannot necessarily be inferred that society made a free, rational, knowledgeable choice, complete with full information, when it allegedly accepted a certain level of risk. Given greater freedom, knowledge, rationality, or monies, past society may well have made quite different 'choices' as to specific levels of risks.

Even if past society had arrived at correct risk decisions, however, it would not follow that those choices ought to be taken as normative for the present or the future. Practitioners of the method of revealed preferences nevertheless make the assumption that one ought to follow risk decisions from the past. In so doing, they subscribe to a highly doubtful presupposition, namely, that values don't change, and that societal norms are not dynamic. But if there is reason to believe that one's knowledge and control of certain risks improve through time, then there may also be reason to believe that present and future standards for the acceptability of those risks ought to be different than they were in the past. In any case, risk acceptability appears to be a function of numerous circumstances, like the degree to which the risk is understood and the extent to which it can be controlled, and not merely a function of what was accepted in the past. If those circumstances change through time, then the acceptability of the risk also is likely to change through time.

From an ethical point of view, one of the most questionable assumptions of the method of revealed preferences is that factors such as probability, magnitude, voluntariness, and resultant economic benefits are sufficient grounds for determining the acceptability of given risks. Obviously, numerous other parameters play a role in the determination of whether a certain risk is acceptable. One of the most important of these is the distribution of a risk and its benefits. In fact, much of the conflict over acceptable risk often arises because the risk or its benefits are inequitably distributed over space, time, or social class. The benefits of a risk may be concentrated, geographically, while the risk is diffuse, or vice versa. With respect to time, the common cases of inequitable distribution are those in which the benefits are immediate but the risks are delayed, as with latent effects of toxic chemicals or radioactive wastes. With respect to social class, a common case is for the benefits to accrue to a particular group, e.g., those who travel by air, but for the costs to be borne by a quite different group, e.g., those who live near airports.

In addition to equity of distribution, a number of other significant parameters are also not accounted for in the method of expressed preferences. Some of these include whether the effect of the risk is immediate or delayed; whether there are available alternatives to taking the risk; whether the exposure to the risk is avoidable; whether the risk is encountered occupationally; whether the relevant hazard is dread or common; whether it affects average persons or merely especially sensitive ones; whether the technology, event, or activity involving the risk is likely to be misused, and whether the consequences of the risk are reversible or irreversible.[33]

Of course, beneath all this discussion of how to evaluate quite diverse risks and benefits, proponents of the method of revealed preferences assume that risks and benefits nevertheless can be known accurately enough in order to arrive at generalizations about the criteria for acceptable levels of safety. This may well be a doubtful assumption, since risks and benefits often have unforeseen second-, third-, and higher-order consequences, and since even first-order consequences are often not known adequately, owing to measurement, modeling, and extrapolation uncertainties.

Another problem with the method of revealed preferences, especially as practiced by persons such as Starr, is that its proponents assume that benefits may be measured either in terms of the money spent by participants in the activity or in terms of the average contribution that the activity makes to one's income. Obviously, however, such expenditures may be good or bad. In using an expenditures criterion for benefits, proponents of the method of revealed preferences fall victim to the classical problems of confusing price with value and preferences with well-being. The discrepancies between price and value, and preferences and well-being, however, are significant because it is rational to assess risks on the basis of the benefits they produce, but only if the alleged benefits are truly connected with human welfare and authentic values. In equating values with what is preferred economically, practitioners of the method of revealed preferences fall into a number of problems. *First*, they ignore the quality of the activities on which they money is spent. This means that the same benefit could be said to accrue to two quite differently valued activities, e.g., using heroin and playing the piano, if it were the case that the money spent by participants in the two activities was the same. *Second*, the equation between values and what is preferred economically blurs the distinction between what makes people good or secures justice and what merely fulfills their wants; in other words, it blurs the distinction between morality and utility. *Third*, the equation between values and economic preferences fails to take account of the fact that wealthy and poor individuals are not equally able to spend funds on a particular activity. It has been shown statistically that, as income increases, people are able to spend more money for environmental quality, medical care, improved life expectancy, home repairs, and job safety.[34] Yet wealthy persons' increased ability to expend more for these amenities does not entail that they value them more than do poorer persons. For example, wealthy persons might be able to spend more on snow skiing than poor persons, but their expenditure for this activity would not mean that skiing were more beneficial than some less costly activity, e.g., bowling, in which lower socioeconomic groups might be more likely to participate. Hence, there are clear discriminatory effects of measuring benefits either on the basis of funds expended on the activity or in terms of the average contribution that the endeavor makes to one's income. If values were determined on the basis of the average contribution that a particular activity, e.g., teaching grammar school, made to the income of the participant, e.g., the teacher, then many activities obviously would be undervalued, while others were overvalued.

There are also some classical economic reasons why it is highly questionable for Starr and other practitioners of the method of revealed preferences to measure benefits in terms of economic expenditures. As was already pointed out in Section 3.3.1, market prices of activities or commodities frequently diverge from authentic values because of the distorting effects of monopoly, the failure to compute externalities, the speculative instabilities of the market, and the absence of monetary-term values for benefits such as natural resources.[35] This means that, because the method of revealed preferences employs criteria for risk acceptability which antecedently presuppose the acceptability of existing market distributions, then those criteria fall victim to the same distortions as do existing market mechanisms.

Of course, the underlying reason why the method of revealed preferences involves so many questionable assumptions is that it faces the same obstacle as the formal method of risk–cost–benefit analysis.[36] Practitioners of both approaches must infer societal values indirectly. Because of the indirectness of these inferences, there may be philosophical and practical objections to them.

3.3.3. *The Method of Expressed Preferences*

One way to circumvent the problem shared by both RCBA and the method of revealed preferences, namely, that their practitioners must infer values indirectly, is to ask people, directly, what risks they deem acceptable. This approach is known as the method of expressed preferences, and it has the obvious merit of eliciting current sentiments about various risks. Advocated by assessors such as Fischhoff, Slovic, and Lichtenstein,[37] this method consists of asking a sample of the public to express its preferences and then analyzing the resulting information. The preferences exhibited in the sample are used to assess the importance of various characteristics of risks (e.g., involuntariness, equity of distribution, etc.) and to rate subjects' perceptions of the risks and benefits accruing to society from various activities, events, and technologies.

Results of the method of expressed preferences show that subjects believe that more beneficial activities may be allowed to have higher levels of risks associated with them. They also indicate that society has a double standard of acceptability for certain hazardous events. For example, two risks may both be involuntarily imposed and may both have the same average annual probability of fatality associated with them. If one risk is catastrophic (likely to kill a large number of people at the same time and at the same place), however, then the public is more averse to it than to similar risks of the same magnitude which are non-catastrophic. Other findings of those who employ the method of expressed preferences are that citizens do not believe that society has managed risk activities so as to allow a higher risk only when a greater benefit is obtained. In other words, practitioners of the method of *expressed* preferences claim to have evidence that, contrary to a basic assumption employed by those who use the method of *revealed* preferences, past societal choices have not determined an optimal risk–benefit trade-off. Rather, say Fischhoff and others, citizens believe that society tolerates a number of activities having high risks and very low benefits, e.g., alcoholic beverages, handguns, motorcycles, and smoking. Moreover, according to practitioners of the method of expressed preferences, surveys indicate that when acceptable levels of safety were compared with perceived benefits, citizens provided evidence that they accepted a risk–benefit relationship much like the one obtained by Starr (see the previous section in this essay). That is, participants in the surveys believed that greater risks should be tolerated only to obtain greater benefits, and that there ought to be a double standard for voluntary and involuntary activities.[38]

Those who employ the method of expressed preferences typically obtain their results through referenda, opinion surveys, detailed questioning of selected groups of citizens, government hearings, and interviewing 'public interest' advocates. Of all these techniques, the survey is probably the most widely used by practitioners of this method. Employment of surveys, however, is saddled with sampling difficulties. For one thing, it is quite difficult to obtain a large sample of individuals with the time and willingness to state their preferences, and the representativeness of any sample group can always be challenged. Some persons may be atypically uninformed or informed, and others may deliberately attempt to bias the survey results. Obviously the success of this method of risk evaluation is in large part a function of the care with which the survey is designed, administered, and monitored.

Another criticism of the method of expressed preferences is that safety questions are too complicated for ordinary citizens to understand. Especially when it comes to new and complex technological issues, people often do not have well articulated opinions. As a consequence, the preferences they express may be highly unstable and incoherent, either because the survey respondents are not familiar with technical terms (e.g., social discount rate) or because their underlying values are incoherent. For example, many persons are highly averse to catastrophic accidents, but are nevertheless willing to fly via commercial carriers. Other instances of incoherent values may arise because the survey respondents play various roles, e.g., those of parent, worker, or citizen, and in each of these roles, they have different preferences about safety. Quite often, citizens may not know even how to think about various risks. For example, they may not feel appreciably different when told that they face an annual risk of death of 10^{-5} as opposed to 10^{-7} from a particular activity. Or, they may not know how to evaluate a probable small increase in cancer risk in the distant future if the activity carrying the risk is one to which they are intensely committed, e.g., use of oral contraceptives. Other difficulties arise regarding the coherence of respondents' survey replies simply because most persons' values change, in some respect, over time and because many people are uncertain as to what should be the basis for their value judgments.

According to Fischhoff, Slovic, and Lichtenstein, at least three features related to the shifting judgments of survey respondents are important. *First*, people may be unaware of changes in their value perspectives and unaware of the degree to which the phrasing of various survey questions elicits particular responses. *Second*, citizens often have no guidelines as to what criteria ought to be used in formulating their value judgments, and which normative perspective is the most desirable one. *Third*, even when citizens do have appropriate guidelines and perspectives, they may not wish to give up their inconsistent ways of valuing safety. All three problems pose severe difficulties for the policymaker attempting to make use of the results of the method of expressed preferences in his public decisionmaking.[39]

3.3.4. *The Method of Natural Standards*

If the greatest flaw in the method of expressed preferences arises from the limitations of the group expressing its preferences, then one way to overcome this difficulty is to have a standard for safety which is independent of the beliefs of a particular society. Instead of examining historically revealed preferences regarding risk or expressed preferences about safety, proponents of the method of natural standards believe that assessors ought to set criteria for risk acceptability on the basis of geological and biological criteria. These geological and biological criteria specify the levels of risk which were current during the evolution of the species. Use of them implicitly presupposes that the optimal level of exposure to various risks is that which has naturally occurred.

One prominent area in which the method of natural standards has been employed is that of setting acceptable levels of radiation exposures. Since normal background levels of radiation average about 170 millirems per year, reasons the U.S. Nuclear Regulatory Commission, this average provides a 'natural standard' against which to measure the acceptability of certain levels of emissions from the U.S. nuclear-reactor fuel cycle. On the basis of this 'natural standard', current annual radiation exposures to the public are required by the U.S. Code of Federal Regulations not to exceed 500 millirems. Such an exposure level is said to be consistent with natural standards since it is roughly of the same order of magnitude as background levels of radiation.[40] This standard, in turn, is a product of perhaps the best known criteria for risk acceptability based on natural standards, those for ionizing radiation as set by the International Commission on Radiological Protection (ICRP).

In setting maximum permissible dose levels for radiation, the ICRP, a small voluntary group, accepts a number of assumptions about safety. *First*, the ICRP presupposes that the natural-standards approach results in only negligible probability of severe genetic or somatic injuries. And *second*, the group presupposes that more frequently occurring effects, e.g., shortened life spans and microscopic changes in one's blood, are either difficult to detect or such that they would be judged acceptable by most persons exposed to them. However, both presuppositions are debatable. Contrary to the first assumption, it could be argued that some natural exposures result in quite severe health effects, e.g., exposure to radon gas from naturally occurring uranium can cause lung cancer; exposure to sunlight can cause skin cancer; ingestion of smoked foods can cause stomach cancer. In other words, even naturally occurring risks, e.g., childbirth, can be quite hazardous, especially if one considers the distant past. The second assumption of the ICRP is equally questionable, since frequently occurring effects, e.g., environmentally induced cancers, do not appear to be judged acceptable by most persons exposed to them. Rather, such risks seem to be tolerated largely because current legal-political mechanisms, e.g., tort law, have not evolved quickly enough to handle technological damages which are either probable, difficult to prove, or only statistical in nature.

As compared to other methods of judging risk acceptability, however, the method of natural standards has several attractive features. One has already been noted; it avoids the limitations of assessment based on societal preferences or behavior. Another advantage is that the method of natural standards avoids converting risks to a common monetary unit, as is usually done in the method of risk–cost–benefit analysis (RCBA) and the method of revealed preferences. It also avoids the problematic tendency to make reference to small probabilities for which most persons have little or no intuitive feelings. Moreover, use of the method of natural standards is likely to produce consistent practices, across hazards, since the same level of acceptability typically is required for the same emission appearing in many different contexts. Another attractive feature of using the method of natural standards is that exposure levels can be set without knowing precise dose–response relationships.

Despite all these advantages, the method of natural standards has a number of drawbacks for which there are no clear solutions. Although there is not time here for an in-depth analysis, some of these drawbacks can be noted. Most obviously, the method fails to take account of benefits. Why should standards be set on a natural basis, in all areas, if one hazard produces great benefits and another does not? A *second* difficulty is that, unless natural exposures diminish (and they seem unlikely in most cases to do so), any new exposure adds to nature's dose and constitutes excess – and perhaps above – 'natural' levels of a pollutant. *Third*, although some technologies, e.g., steelmaking, produce many pollutants, each of which constitutes a small and naturally acceptable effluent, the synergistic effects of all these small levels of different compounds are not addressed by the method of natural standards, even though they may be responsible for dangerous consequences. *Fourth*, the method provides no basis for assessing the value of making trade-offs of risks, e.g., replacing dirtier technologies with cleaner ones, or replacing one type of risk associated with a particular technology with an allegedly smaller risk (associated with the same technology) which causes higher risks in other areas or to other people. As an example of the second type of risk replacement, consider the case of removing wooden guard rails from workplaces in automated meat-cutting industries, so as to reduce the risk of meat infection induced by unsanitary guard rails. Although removal of the rails might reduce the consumers' risk of obtaining infected meat, it very likely increases the risk of worker injury because of their unprotected exposure to large, meat-cutting machinery. Since the approach provides grounds only for judging individual increases in pollutants, it is not in accord with some of our best intuitions about how to increase safety. *Fifth*, for some new substances, there is no historical tolerance for them. If one followed the natural-standards approach in such cases, e.g., that of saccharin, then one would have to propose exposure levels which tolerated none of the substance at all, regardless of how beneficial to health it might be. *Sixth*, following the natural-standards approach, one might be inclined to say that certain naturally occurring levels of pollutants causing a given number of annual fatalities were acceptable. If one reasoned in this manner, following the method might cause him to view certain numbers of fatalities as tolerable, even though they might be avoidable through modern technology. Hence use of this method might encourage policymakers to be content with the *status quo*, rather than to seek progress in risk management and control. Such a consequence appears undesirable, especially in more developed countries. This is because, as personal wealth increases, one is likely to demand higher and higher standards for health and safety. In other words, some interpretations of the method might cause one to be too liberal in accepting preventable fatalities allegedly resulting from acceptance of the natural-standards approach.[41]

4. CONCLUSION

The real problem with the method of natural standards, however, is not that adherence to it leads one to accept, erroneously, high numbers of fatalities. Rather, the real difficulty with this method, as well as with the other approaches to risk evaluation (RCBA, revealed preferences, and expressed preferences), is that assessors often forget the methodological assumptions which limit the validity of their risk-evaluation conclusions. In other words, the real difficulty is not that each of the three stages of risk assessment (risk identification, risk estimation, and risk evaluation) involves methodological assumptions but that, in practice, these assumptions are often ignored. As a consequence, risk assessment results are often viewed as far more objective than they really are. This, in turn, means that policy conclusions based on the assessment results are frequently more controversial and value-laden than is thought.

If risk policy is more value-laden than is recognized, then the task of the philosopher approaching risk assessment is clear. He ought to uncover the ethical and methodological commitments implicit in risk-assessment techniques and subject them to explicit analysis. In thus rendering them explicit, he will increase the probability that authentic values will dictate our life-and-death decisions, rather than that our philosophy will be used to rationalize our public policies.

NOTES

1 W. J. Librizzi, 'Love Canal', in *Risk in the Technological Society* (ed. by C. Hohenemser and J. Kasperson), American Association for the Advancement of Science and Westview Press, Boulder, 1982, pp. 61–76; hereafter cited as: Hohenemser and Kasperson, *Risk*. See also L. Ember, 'Love Canal', in Hohenemser and Kasperson, *Risk*, pp. 77–102.

2 See, for example, R. Peterson, 'Three Mile Island', in Hohenemser and Kasperson, *Risk*, p. 35.

3 See K. Shrader-Frechette, *Nuclear Power and Public Policy*, 2nd edition, Reidel, Boston, 1983, pp. 88–89, 97–98; hereafter cited as: *Nuclear Power*.

[4] C. Hohenemser and J. Kasperson, 'Overview', in Hohenemser and Kasperson, *Risk*, p. 17.

[5] U.S. Nuclear Regulatory Commission, *Reactor Safety Study – An Assessment of Accident Risks in U.S. Commercial Nuclear Power Plants*, Report No. (NUREG–751014) WASH-1400, Government Printing Office, Washington, D.C., 1975, is the famous 'Rasmussen Report'. See Appendix XI for the critique of the Union of Concerned Scientists and that of other scientific bodies. For the APS critique, see H. W. Lewis, *et al.*, 'Report to the American Physical Society by the Study Group on Light-Water Reactor Safety', *Reviews of Modern Physics* **XLVII** (1), (Summer 1975), S1–S124. See also C. Hohenemser, R. Kasperson, and R. Kates, 'The Distrust of Nuclear Power', *Science* **CXCVI** (4285), (April 1977), 25–34. See Shrader-Frechette, *Nuclear Power*, pp. 78–90, for a discussion of these points and for more bibliographic information regarding the relevant risk assessments of nuclear power and criticisms of them.

[6] R. W. Kates, *Risk Assessment of Environmental Hazard*, Wiley, New York, 1978, p. 46; hereafter cited as: Kates, *RA*.

[7] See, for example, W. Lowrance, *Of Acceptable Risk*, Kaufmann, Los Altos, California, 1976, pp. 70–74; hereafter cited as: Lowrance, *OAS*.

[8] See C. Starr, 'General Philosophy of Risk–Benefit Analysis', in *Energy and the Environment* (ed. by H. Ashby, R. Rudman, and C. Whipple), Pergamon, New York, 1976, pp. 28–30; hereafter cited as: Ashby, *et al.*, *EAE*.

[9] See C. Starr, 'Social Benefit versus Technological Risk', *Science*, **165** (3899), (19 September 1969), 1232–1238; hereafter cited as: Benefit.

[10] See L. Lave, 'Methods of Risk Assessment', in *Quantiative Risk Assessment in Regulation* (ed. by L. Lave), Brookings Institution, Washington, D.C., 1982, pp. 28–29; hereafter cited as: Methods and *QRA*.

[11] Lave, Methods, p. 29, uses this example.

[12] Lave, Methods, p. 30, and Frank Press, *Risk Assessment in the Federal Government: Managing the Process*, National Academy Press, Washington, D.C., 1983, p. 23; hereafter cited as: *RA*.

[13] Lave, Methods, p. 30, and Press, *RA*, pp. 22–23.

[14] Lave, Methods, p. 30, and Press, *RA*, p. 22.

[15] E. Lawless, M. Jones, and R. Jones, *Comparative Risk Assessment*, Draft Final Report, Grant No. PRA-8018868, National Science Foundation, Washington, D.C., 1983, pp. 118–119; hereafter cited as: *CRA*.

[16] Press, *RA*, pp. 19–20.

[17] Press, *RA*, pp. 23–27.

[18] Lave, Methods, p. 39, and Press, *RA*, p. 24.

[19] See Press, *RA*, pp. 24–25, and Lawless, Jones, and Jones, *CRA*, pp. 121–124.

[20] See Press, *RA*, p. 24, and Lawless, Jones, and Jones, *CRA*, pp. 121–124.

21 See Lave, Methods, pp. 49–54; Press, *RA*, pp. 27–28; and E. Crouch and R. Wilson, *Risk/Benefit Analysis*, Ballinger, Cambridge, Massachusetts, 1982, pp. 51–73; hereafter cited as: Crouch and Wilson, *RBA*.

22 B. Fishhoff, S. Lichtenstein, P. Slovic, S. Darby, and R. Keeney, *Acceptable Risk*, Cambridge University Press, Cambridge, 1981, p. 101; hereafter cited as: *AR*.

23 For a history and overview of RCBA, see K. Shrader-Frechette, *Science Policy, Ethics, and Economic Methodology*, Reidel, Boston, 1984, esp. Chs. 1–2; hereafter cited as: *Science Policy*.

24 For analysis of these and other criticisms of RCBA, see Shrader-Frechette, *Science Policy*, esp. Chs. 5–7.

25 See the previous note.

26 For expressions of this point of view, see S. Dreyfus, 'Formal Models vs. Human Situational Understanding', *Technology and People* **1** (1982), 133–165, and R. Socolow, 'Failures of Discourse', in *Ethics and the Environment* (ed. by D. Scherer and T. Attig), Prentice-Hall, Englewood Cliffs, 1983, pp. 139–151.

27 For a careful defense of analytic decision methods for risk evaluation, see Shrader-Frechette, *Science Policy*, Ch. 2.

28 See C. Starr, 'General Philosophy of Risk–Benefit Analysis', in Ashby, *et al.*, *EAE*, p. 6, and Starr, Benefit, pp. 1235–1236.

29 For examples of such tables, see Fischhoff *et al.*, *AR*, pp. 81–83, and B. I. Cohen and I. Lee, 'A Catalog of Risks', *Health Physics* **36** (6), (June 1979), 708–721; hereafter cited as: Catalog.

30 Cohen and Lee, Catalog, p. 720.

31 Starr, Benefit, p. 1237.

32 See H. J. Otway and J. Cohen, 'Revealed Preferences: Comments on the Starr Benefit–Risk Relationships', IIASA RM 75–5, International Institute for Applied Systems Analysis, Laxenburg, Austria, March 1975.

33 See Lowrance, *OAS*, pp. 86–94.

34 P. S. Albin, 'Economic Values and the Value of Human Life', in *Human Values and Economic Policy* (ed. S. Hook), New York University Press, New York, 1967, p. 97, and M. Jones-Lee, *The Value of Life: An Economic Analysis*, University of Chicago Press, Chicago, 1976, pp. 20–55.

35 See note 24.

36 For evaluation of the method of revealed preferences, see Crouch and Wilson, *RBA*, pp. 75–78; Fischhoff, *et al.*, *AR*, Chapters 5 and 7; and W. D. Rowe, *An Anatomy of Risk*, Wiley, New York, 1977, pp. 79–80, 259–359.

37 See B. Fischhoff, P. Slovic, and S. Lichtenstein, 'Weighing the Risks', *Environment* **21** (4), (May 1979), pp. 32–34; hereafter cited as: Weighing. See also the Fischhoff *et al.* book and articles listed in the bibliography of this volume.

[38] Fischhoff, *et al.*, Weighing, pp. 32–33.
[39] Fischhoff, *et al.*, Weighing, pp. 32–33.
[40] *Code of Federal Regulations*, 10, Part 20, U.S. Government Printing Office, Washington, D.C., 1978, p. 189.
[41] Fischhoff, *et al.*, AR, pp. 87–88.

PART TWO

ETHICAL PROBLEMS WITH THE METHOD OF REVEALED PREFERENCES

CHAPTER THREE

THE COMMENSURABILITY PRESUPPOSITION

1. INTRODUCTION: A LESSON IN METHODOLOGY

Ten years ago, it was commonplace for economists to calculate the value of human life as the lost economic productivity associated with a shortened life span.[1] As is well known, such a view has been shown to be grossly inadequate. Most obviously, it leads to counterintuitive results, such that the value of the life of a 65-year-old laborer is equal to the sum of his remaining earnings until retirement or that the value of the life of a small child is near zero, since her future earnings are discounted at a market rate of interest.

From a methodological point of view, the great tragedy of using this formula, to estimate health risks and benefits, is not merely that it misrepresents the worth of human life. A significant problem is also that uncritical use of such simplistic assumptions has probably helped to produce a reactionary rejection of *all* attempts, even sensitive and sophisticated ones, to analyze rationally certain risks and benefits and to quantify many parameters of various safety programs, two necessary components of reasoned policymaking. My purpose here, however, is not to outline a defense of analytic assessment techniques, even though I strongly support them.[2] My concern, instead, is to argue that these analytic methods ought to be improved. In particular, I argue that assessors ought to reject uncritical use of one highly doubtful methodological tenet which I call the 'commensurability presupposition'. This is the assumption that the marginal cost of saving lives, across opportunities, ought to be the same. If this presupposition is not used more cautiously, then its employment may jeopardize both the success and the acceptance of analytical risk assessment, much as incautious use of the earlier presupposition (that human lives ought to be valued solely in terms of remaining discounted economic productivity) has jeopardized both the success and the acceptance of cost–benefit analysis.[3]

In this essay, I examine some of the doubtful epistemological and ethical premises upon which the commensurability presupposition rests. In general, I ask whether subscribing to this presupposition commits one to begging the question of whether a given level of risk is acceptable. In particular, I analyze three claims often made to support the presupposition. These are the equity claim, the rationality claim, and the responsibility claim. Next I evaluate several criteria in terms of which the use of the commensurability presupposition may be judged acceptable or unacceptable in certain situations. I maintain that, while use of this presupposition is acceptable in specific types of 'easy cases', it is unacceptable in so-called 'hard cases'. I argue that, although the commensurability presupposition ought not to be accepted in all instances, two principles help to determine the cases in which its use is acceptable. These are the principle of *prima facie* egalitarianism and the principle of everyone's advantage.

2. THE COMMENSURABILITY PRESUPPOSITION AND CURRENT RISK ANALYSIS

Faced with the task of evaluating the acceptability of various risks, analytic assessors usually address at least two questions. (1) How has society valued certain risks? And (2) given these evaluations, how ought public monies be spent to reduce various societal risks? Assessors often employ the commensurability presupposition because it provides them with a clear criterion for answering question (2). Provided that use of this criterion is consistent with acceptable societal evaluations, it defines an expenditure for risk reduction as desirable, so long as the same funds could not be used more cost-effectively to reduce a greater risk.

In attempting to answer question (1), assessors generally follow one of two methodologies "for inferring social values": "revealed preferences" or "observed preferences",[4] and "expressed preferences" or "psychometric survey".[5] Apart from which of these two methodologies they follow in answering question (1), risk assessors generally adhere to the commensurability presupposition when they address question (2). Theorists such as Starr, Whipple, Hushon, Okrent, Maxey, Cohen and Lee, for example, agree that the effort required to control societal risk, as measured by the cost per life saved, ought not to vary from one risk to another. Moreover, they claim, preferred safety policies ought to be those which save the greatest number of lives for the least amount of money.[6]

In defending the commensurability presupposition, its proponents typically rely on three different arguments, each of which deserves careful analysis. I call these the equity claim, the rationality claim, and the responsibility claim.

2.1. *The 'Equity Argument' for the Presupposition*

The *equity claim* is that policymakers ought to follow the commensurability presupposition because doing so will provide all people with equal treatment and equal protection from societal hazards. Advocates of politically 'left' policy, as well as risk assessors, often champion this claim. They maintain that, unless the commensurability presupposition is followed, more funds will be spent to safeguard the health and safety of politically powerful groups, while fewer monies will be spent to protect the well-being of politically powerless people.

In defending the equity claim, one critic of risk assessment asks, for example, why government regulations on coke-oven emissions have been designed "to protect the lives of steelworkers at $5 million each", while a national Pap smear screening program, "that would save women's lives at less than $100 thousand each", has gone unfunded.[7] Similar equity-related questions have been raised regarding other societal risks, particularly in cases where government appears to be reducing the risks faced by the rich and powerful more than those borne by ordinary people. Airplane risks, for example, are faced primarily by those who have above-average wealth and political power, while automobile risks cut across a wider and more diverse group of affected persons. According to some theorists, society spends more to save people from airplane accidents than from automobile accidents precisely because airline safety is championed by the powerful people most likely to be affected by it. On this view, the common man has no comparable champion. In the absence of a societal mandate to follow the commensurability presupposition, say proponents of the equity claim, the net result is that societal expenditures implicitly value the lives of air travelers more than those of automobile travelers.

After noting that in France $30,000 is spent annually per life saved through automobile accident prevention, while $1 million is spent annually per life saved through airplane accident prevention, Okrent criticizes the fact that the same "value of life" was not used.[8] Likewise, Starr and Whipple maintain that, in an *optimum* safety policy, "the comparative marginal cost-effectiveness of each opportunity for saving lives would become the guiding principle in the allocation of resources, and the value of life would be implicit in the total national allocation of funds."[9] According to participants in a recent panel on risk assessment, use of the commensurability presupposition is one way to treat people equitably; "in terms of establishment of uniform rates of the value of life, a basic unit for health insult is needed. Different agencies must reach a common basis for making decisions."[10] If they do not, claim assessors, our cherished traditions of equal protection will be eroded.

2.2. *The 'Rationality Argument' for the Presupposition*

Proponents of the *rationality claim* make a somewhat different case for the commensurability presupposition. In appealing for consistency in the marginal cost of risk reduction across opportunities, they assert that their goal is "to make people approach risks more rationally".[11] Supporters of this argument appear to believe that there is a univocal concept of rationality. *Rational* choices, for them, are *economically efficient* choices. For proponents of the rationality argument, like Häfele and Okrent, subscribing to this univocal concept of rationality is merely a matter of being consistent. "Consistent", rational people, they claim, would spend the same amount of money for the same level of risk abatement, regardless of the risks involved. People do not deal with all hazards in a consistent and rational manner, says Häfele, because they have unreasonable fears stemming largely from their ignorance of technology, especially new technology. He maintains that they have irrational perceptions of how to deal with risks such as those from liquefied natural gas (LNG) and nuclear facilities. Following this same line of argument, Okrent asks why we spend great sums of money to save persons from accidents at LNG facilities, but virtually nothing to save them from natural disasters such as flooding.[12]

Fischoff, Slovic, and Lichtenstein ask why our legal statutes are "inconsistent", why are they less tolerant of carcinogens in the food we eat than of those in the water we drink or the air we breathe? Why, they ponder, should we spend more to protect people from carcinogens in food than from carcinogens in air or water? In the United Kingdom, they maintain that this same inconsistency is apparent. 2,500 times as much money is spent there, to save one life through safety measures in the pharmaceutical industry, as is spent to save one life through safety measures in the agricultural industry.[13]

For proponents of the rationality claim, failure to subscribe to the commensurability presupposition is to be guilty of inconsistency. They believe that it is not consistent to expend funds to achieve a particular level of health and safety in one area, if spending the same or fewer monies would provide a greater level of health and safety in another area. In other words, the key assumption behind the rationality claim is that reasonable, consistent people save the 'cheapest' lives first, or spend funds so as to maximize health and safety for the greatest number of people.

2.3. *The 'Responsibility Argument' for the Presupposition*

Another defense of the commensurability presupposition is made on the grounds of the *responsibility claim*. Proponents of this view argue that, if one does not follow the commensurability presupposition, then he is responsible for the lives or health lost by pursuing schemes which do not save the 'cheapest' lives first. They maintain, for example, that if one opts to spend $1 million to save one life through airplane accident prevention, when the same expenditure for automobile accident prevention could have saved 33 lives, then one is responsible for the 32 excess deaths.[14]

Proponents of the responsibility claim place great weight on the fact that resources for the reduction of risk to the public are not infinite. They maintain that if we tolerate different marginal costs for saving lives, "if we are spending the available resources in a way that is not cost-effective, we are, in effect, killing people" and are "responsible for unnecessary deaths".[15]

By virtue of their emphasis on spending finite funds so as to save the greatest number of lives, proponents of the responsibility claim implicitly appeal to a triage way of thinking. Borrowed from French military medicine, the term "triage" refers to doctors' practice of separating battlefield wounded into three groups: (1) those for whom immediate medical care will make a difference between life or death; (2) those who will likely die, even if they receive immediate medical attention; and (3) those who will likely live, whether or not they receive immediate medical attention. Since the doctors have limited resources (time and medical supplies), they are often said to have a duty to attend first to the wounded in category (1). If they do not follow this strategy, they might be said to be responsible for excess battlefield deaths. For the proponents of the responsibility claim, limited medical resources on the battlefield are analogous to the limited societal resources for risk abatement. They believe that, just as the battlefield doctors ought to maximize the numbers of lives saved with their finite resources, so also policymakers ought to follow the commensurability presupposition and maximize the number of deaths averted through risk abatement. Otherwise, they are 'responsible' for excess deaths.

3. REASSESSING THE COMMENSURABILITY PRESUPPOSITION

3.1. *Problems with the Presupposition*

Plausible as are these three appeals for equity, rationality, and social responsibility, there are a number of reasons why none of them provides convincing grounds for adherence to the commensurability presupposition in all, or even most, instances. Let us examine each of the claims more closely.

3.1.1. *Five Faulty Assumptions Underlying the Equity Claim*

The *equity claim* rests on the fundamental assumption that, to guarantee persons *equal* protection, the marginal costs of saving lives ought to be the *same* for all risks. In other words, proponents of this view maintain that *sameness* of expenditures (i.e., spending the same amount per life saved), across risk reduction schemes, is necessary for providing persons with *equality* (sameness) of protection.

This is a powerful claim and a compelling assumption, especially because many people chafe at the thought of *different* amounts of money being spent to save persons' lives, depending on whether they are rich or powerful or male, precisely because they are rich or powerful or male. Our populist sympathies probably lie almost totally with sameness of expenditures. Discrimination in protection is profoundly disturbing. Yet, the *emotive* power of the equity claim camouflages a number of reasons why it fails to provide convincing *ethical* grounds for following the commensurability presupposition. In the following paragraphs, I will uncover the precise nature of the errors inherent in the equity claim. These errors include the assumption that sameness of risk expenditures provides either necessary or sufficient conditions for sameness of protection against risk; that sameness of protection is not significantly different from equality of protection; that sameness of concern or respect is not significantly different from sameness of treatment, and that rights to equality or sameness of protection are rights in the strong sense. Let us examine each of these assumptions in detail.

In arguing that the marginal cost of saving lives, across public safety programs, ought to be the same, proponents of the equity claim appear to be assuming that persons subject to diverse risks all need the same level of protection from the government. It is not clear, however, that everyone needs the same level of expenditures to protect him from given societal risks. Some proponents of the equity claim maintain, for example, that policymakers who pursue risk reduction ought to cost the lives of automobile travelers the same as those who travel by plane (see note 8). However, the safety of automobile travelers is far more dependent on individual choices, e.g., whether to ride with a drunken driver, or whether to use a seat belt, than is the safety of airplane passengers. In this sense, it is not clear that automobile travelers *need* the same level of government-guaranteed risk reduction as airplane passengers do. If the airplane and auto cases are fundamentally disanalogous in that the individual has more control over risk reduction in automobiles, then it is less plausible to argue that the government ought to guarantee the same levels of expenditures for both, thereby ignoring the effects of an individual's behavior on his level of protection.[16] At least in the case where the same expenditures are not needed to provide the same protection, sameness of expenditures is not a necessary condition for the same protection.

Second, not only is sameness of expenditures not a necessary condition for the same protection against various risks, but it also is not a sufficient condition. Simply costing people's lives the same, across different risk reduction programs, does not guarantee them the same protection. If one costed lives the same, for example, in reducing auto risk and radiation risk, this would not provide the same protection for all potential victims of automobile accidents and radiation exposures. In part this is because medical differences among persons place some of them at a higher risk from radiation. Children, those with allergies, and those with previous x-ray exposures, bear a higher risk than do others when all lives are costed the same. Providing the same level of protection is impossible, given persons' different levels of susceptibility, but the same expenditures for risk reduction.[17]

Third, even if the same expenditures (across safety programs) did guarantee the same level of protection from risk, it is not clear that there are always morally relevant reasons why persons ought to receive the same protection. Genuinely *equal* protecton might require protection which is not the *same* for all individuals. Without going into lengthy consideration of why all individuals might not deserve the same protection in all situations, I can suggest a few cases in which, given appropriate conditions, one ought to allow exposure to different levels of risk. Although these are not universalizable, the cases might include circumstances in which the different protection is given/allowed:

(1) as a reward for merit or virtue;
(2) as a recompense for past deeds or actions;
(3) as an incentive for future actions held to be highly socially desirable; or
(4) as a provision for special needs.

Obviously, for example, if one wishes society to have the benefit of the services of those persons who perform much needed functions, e.g., acting as President of the United States, then one must somehow give those persons better than equal protection in certain respects. Following circumstances (3) and (4), this 'better' protection might be justified on grounds of incentives necessary to attract candidates to the office of President or necessary to provide stability to the country.

As legal philosopher Ronald Dworkin observes, there do not appear to be ethical grounds for claiming that everyone ought to receive the same treatment. There appear to be ethical grounds for claiming only that everyone ought to have the same *concern* or *respect* in the political decision about how goods, protection, and opportunities are to be distributed.[18] His point, and my observation about circumstances (1)–(4) above, is not that anyone's rights may be ignored in the safety calculations, but that one's interests may be outweighed by another's interests. For example, in certain circumstances, protecting the President of the U.S. may outweigh protecting a particular citizen. If this is so, then one ought to choose the risk-abatement policy which gives everyone the *same* concern or respect, but which provides an equitable basis for deciding when one person's interests outweigh another's. This means that use of different marginal costs, per life saved across opportunities, cannot be shown to be unethical merely because its effects do not give persons the same protection. To show this, one must argue either that there were no morally relevant reasons for different protection, or that persons were not given the same concern or respect, or that some persons' interests were erroneously judged to outweigh those of others.

A fourth and final reason why one might not argue either for sameness of risk-reduction expenditures, or for sameness of protection in risk situations, is that we may not have a right to equal protection in what Dworkin calls "the strong sense". Instead, it may only be a right "in the weak sense". The strong and weak senses of rights are distinguished, according to Dworkin, because offenses against strong rights and weak rights are different from each other in *character*, and not just in degree. Offenses against weak rights limit only one's liberty, says Dworkin, but assaults on strong rights limit one in a way which goes beyond liberty. These assaults constrain values or interests, necessary to protect either one's dignity or standing as equally entitled to concern and respect. In other words, liberties protected by strong rights are defined by other criteria (e.g., dignity, security) in addition to liberty, whereas liberties protected by weak rights are defined only by the criterion of liberty. For Dworkin, rights in the strong sense may never be denied by government, even when it is in the general interest to do so. Rights in the weak sense may be denied, however, when it is in the general interest.

For example, says Dworkin, I have no right in the *strong sense* to drive down a particular street, even if I have a driver's license.

This is because my right to drive down the street is not necessary to protect my dignity or standing. Hence if it is in the general interest to make some street one way, then my liberty to drive down some streets, in either direction, is denied. For this reason, one has a right to drive down a particular street, but only in a weak sense. Because it is necessary to protect one's dignity and standing, however, Dworkin claims that we do have a right *in the strong sense* to free speech, for example. (See notes 16–18 to this chapter.)

If indeed Dworkin's distinction between strong and weak rights is correct (a distinction which I don't wish to take the time to defend here), and if rights to equal (the same) protection are rights only in a weak sense, then all persons ought not receive the same protection in all situations. But if so, then it is not obvious that the equity claim (which demands the same protection in all situations) provides support for the commensurability presupposition. To support this presupposition, proponents of the equity claim ought, in addition, to show that there are *good reasons* (in a particular case) to spend the same funds per life saved across opportunities. In other words, proponents of the equity claim make too simplistic an appeal to equal protection. At best, they are able to show only that, *in a given case*, good reasons support spending the same funds per life saved. Later in the essay (Sections 3.2 and 3.3), I will discuss what some of these good reasons might be.

3.1.2. *Two Faulty Assumptions Underlying the Rationality Claim*

Equity, however, is not the only basis alleged to support the commensurability presupposition. Numerous assessors have argued for it on the grounds of the *rationality claim*. This is the assertion that reasonable, consistent people would agree to spend funds in such a way as to maximize health and safety for the greatest number of people. In other words, they assume that rational people would choose to spend monies on risk reduction so as to save the greatest number of lives, or buy the greatest amount of health for the available dollars. However, there are a number of problems with this line of argumentation. What Okrent and other assessors forget, when they appeal to the rationality of setting risk reduction priorities on the basis of economic efficiency, is that economic efficiency is neither the sole, nor necessarily the best, criterion for rational policymaking. There are no common denominators among the costs, risks, and benefits of various health and safety programs. As a consequence, numerous evaluational assumptions are required in order to *define* some parameter as a basis for comparison among alternative programs. In this case, many assessors assume that 'cost per life saved' provides an adequate basis in terms of which safety options may be ranked, even though this parameter alone is clearly not sufficient grounds for preferring one program over another. That it is not sufficient grounds is shown by numerous examples of rational decisionmaking not based on 'cost per life saved'.

Consider four cases, in each of which safety options are ranked by a criterion other than economic efficiency. In the first, the so-called 'rights case', the goal of guaranteeing rights trumps the goal of economic efficiency. Suppose, for example, the government funds two safety programs, one to protect seasonal campers from volcanic activity in a given region, and one to protect civilians from radioactive emissions at a government weapons-testing facility. Obviously, government has a greater responsibility to protect citizens from risks which it involuntarily imposes than to protect them from those which are solely natural, or which they impose on themselves by their decision to camp in a particular area. This is in part because radioactive emissions can be controlled or mitigated, while many natural disasters cannot. Government therefore has more responsibility for that which is under its control. Second, the Fifth and Fourteenth Amendments to the U.S. Constitution provide guarantees against *state* action which limits equal protection, not against all natural events threatening well-being. This means that, although one has a legal right to equal protection from state-induced hazards, one has no clear legal right to equal protection from natural hazards. This being so, risk reduction priorities in the volcanic safety/weapons safety case need not be set according to economic efficiency. Even though it might cost *more* to save lives/health in the weapons safety case, this might be the more rational policy choice. The reason is that victims in the weapons situation might have a stronger legal claim for rights violations than would victims in the volcano case.

In a second, or so-called 'civil liberties case', the goal of maximizing civil liberties might trump the goal of economic efficiency. Numerous examples of this situation come to mind. Suppose, for instance, that the government is funding two safety programs, one to provide centers for voluntary, no-cost checking of blood pressure, and the other to monitor drunken driving by administering mandatory breath tests, between 10 p.m. and 6 a.m., to all drivers at selected checkpoints. Even if it could be shown that more lives per dollar would be saved by checking late-hour drivers, rather than by providing free blood-pressure monitoring, this would not mean that additional expenditures for administering breath tests were more desirable. A rational person or government, admittedly opposed to drunk driving, could well decide to maximize civil liberties and to avoid coercive controls on such driving. His decision to let civil liberties 'trump' economic efficiency might be justified in part on the grounds that citizen participation in blood pressure checks would be voluntary, while monitoring late-hour drivers would be involuntarily imposed.

Third, people or government could well make a rational decision to let pleasure trump economic efficiency. For example, suppose government were trying to decide whether to spend funds to prohibit the addition of sugar to selected cereals, breads, and juices, or to use the same monies for mass screenings for TB. Even if it could be shown that more lives per dollar could be saved by regulating sugar, it is not clear that government ought to accord this a higher priority than TB screenings. This is because a rational policy might allow people to maximize their pleasure from eating certain foods, rather than simply to save the greatest number of lives. Obviously many people choose, on given occasions, to maximize their pleasure rather than their health or safety. To claim that such valuations are irrational is to assume, erroneously, that the only rational, human goal is to protect health and safety. However, without some pleasure, perhaps health or safety is of little concern. As Bergstrom put it, "I know of no one who acts as if he subordinates all desire for other pleasures to maximizing his probability of survival."[19] After all, those who engage in hang gliding, sky diving, motorcycle riding, or even in eating rich desserts or drinking saccharin-spiked soda, can hardly be said to be making irrational choices.

Fourth, rational policymaking could well be based on maximizing temporal, rather than economic, efficiency. For example, each time a person decides to take a trip on a small, private plane rather than on a commercial carrier, because the former mode of travel is quicker, he is maximizing temporal, rather than economic, efficiency. Analogously, the government could well decide rationally to spend funds in such a way that would enable people to save time, rather than to maximize health/safety, especially if the risks to health and safety were very small and the time gains were very great. For example, the government might decide to spend safety funds to regulate automatic doors on subways rather than to spend the same monies to mandate speed reductions on the same trains. Even if the speed-regulation programs saved more lives per dollar than did the door-safety programs, it could still be argued that efficiencies in time are sometimes more important than economic efficiencies and that the monies ought to be spent to regulate the doors.

What these four cases illustrate is that 'cost per life saved' is often not a sufficient criterion for ranking public policy options. Because it is not, proponents of the rationality claim err in valuing health and safety above all else. Every time that a society affirms that it is willing to go to war, for example, it assents to the thesis that it values things other than health and safety. On a much more mundane level, people daily make trade-offs in favor of other kinds of values over safety: excitement, comfort, freedom, for instance. Many people also choose to live in a city, even though it is by now evident that doing so threatens their health and safety more than would rural life. In other words, safety or security in itself is not always something of value. Whether it is or not depends on what it and the threat to it are, and on what ends it is meant to serve. Security is merely the assurance that given interests will not be adversely affected. Since most policies decrease the ability to meet one set of interests, but increase the ability to fulfill others, no policy can be viewed as desirable purely on the grounds of meeting the interests of security.[20]

The insufficiency of the cost-per-life-saved criterion illustrates very well what Shackle calls "the problem of the single maximand".[21] There is no *single* parameter according to which two different safety programs may be ranked. This is because problems of resource allocation require multivariate solutions, taking many variables into account, rather than maximizing only the health or safety factor. As he puts it, society has no "weighing machine for the value of two actions".[22]

Besides the fact that there is no single maximand in deciding policy regarding health and safety, there is another problem with the rationality claim. This is that proponents of the claim are misguided in appealing to the *consistency* of the costs per life saved across opportunities. The appeal is misguided because allocative decisions in no way place a 'value' on human life. If government spends x dolloars to save y lives in one case, but w dollars to save z lives in another case, no inference about the relative value of life, in the two cases, can be drawn, purely on the basis of the relative expenditures. It is not true that the value of life in the first case is $\$x/y$, but that it is $\$w/z$ in the second case. Rather, the expenditures merely reveal a particular decision about how policymakers believe government ought to spend a given sum of money. More funds (per life saved) may be spent to aid victims in one situation, rather than another, because some victims are less able to control their own health/safety, or because certain classes of victims have a greater *right* to protection, or for some other reasons. Because such expenditures may be justified by appeal to many different value judgments, they do not necessarily contain any implicit statements about the relative value *of life* in various cases.[23] Hence it makes no sense to appeal to the *consistency* of 'valuations of life' across opportunities. Moreover, appeals for consistency of expenditures per life saved ignore the fact that many cases are quite disanalogous. One case might concern an involuntarily imposed, catastrophic risk to health and safety, for example, while another might concern a voluntarily chosen, individual risk. If the risks are dissimilar in these and/or other relevant respects, it is not clear why they should be treated consistently with respect to expenditures per life saved. Hence it is not clear that one ought to support the rationality claim by a simple appeal to consistency across cases. Moreover, as the problem of the single maximand and my four counterexamples show, rational policymaking cannot be identified simply with maximizing economic efficiency. But if it cannot, then the rationality claim does not provide unequivocal support for the commensurability presupposition. At best, proponents of this claim are able to show only that, in a given case, good reasons support spending equal funds per life saved. Later in the essay, I will discuss what some of these good reasons might be.

3.1.3. *Two Faulty Assumptions Underlying the Responsibility Claim*

Just as there are problems with using the rationality claim and the equity claim to support the commensurability presupposition, so also there are difficulties with the responsibility claim. Formulated by Okrent and others, this claim is that whoever tolerates different marginal costs for saving lives is "responsible for unnecessary deaths".[24] In other words, if societal monies are not spend so as to save the greatest number of lives per dollar, then those who made this fiscal decision are responsible for the excess deaths, just as a battlefield doctor who does not follow triage is responsible for excess casualties. To avoid responsibility for having killed people, goes the argument, one ought to support the commensurability presupposition (see Section 2.3 earlier).

The plausibility of the responsibility claim is a function both of its intuitive appeal and its implicit use of the economists' notion of opportunity costs. On the intuitive level, numerous people would probably want to say that policymakers who use tax monies to build bombs, rather than to feed impoverished people, are responsible for the malnourishment of many persons. They would probably want to claim that the bomb builders are responsible for exacerbating poverty, much as Okrent wants to claim that those who reject the commensurability presupposition are responsible for 'killing people'.

Such claims of responsibility are also plausible in the light of the fact that persons are routinely held responsible for the opportunity costs of their actions and that excess deaths may be considered to be one type of such a cost. Every expenditure for a given program has (what an economist would call) an 'opportunity cost'. This is defined as the cost of foregoing other options for investment or for spending the funds. The opportunity costs of funds used in the production of good *A* consist of foregone opportunities of producing good *B*. For example, using water as a receptive medium for pollutants has, as an opportunity cost, foregoing use of the water for drinking.[25]

Just as an industry is responsible for the opportunity costs of its decision to use water as a receptacle for pollutants, perhaps also the policymaker could be said to be responsible for the opportunity costs of his decision to fund one risk abatement program rather than another. And if funding this program causes fewer lives per dollar to be saved, then the policymaker might be said to be responsible also for the excess deaths, since these are part of the opportunity costs of his decision.

Proponents of the responsibility claim – who maintain that policymakers are responsible for the 'excess deaths' caused by pursuing one safety program over another – err, in general, by failing to limit the scope of their concept of responsibility. In subscribing to a notion of responsibility which is virtually unbounded, they make two main assumptions, both of which deserve further discussion. The first assumption is that one's responsibility for the opportunity costs of his actions always requires him to avoid the opportunity costs of excess deaths. The second assumption is that because excess deaths are allegedly an opportunity cost of one's actions, one is clearly responsible for them. Several considerations reveal that both assumptions are doubtful.

3.1.3.1. *The assumption that one ought to avoid the opportunity cost of excess deaths.* Consider first the assumption that one always ought to avoid the contingent opportunity cost of excess deaths. In making this assumption, proponents of the responsibility claim appear to believe that the opportunity costs of various safety investment programs are analogous and ultimately reducible to issues of health and survival. On the contrary, they do not seem to me to be reducible in so simple a fashion.

The opportunity costs of two investments often differ because their *purposes* differ. The two programs mentioned by Okrent, for example, prevention of deaths from canyon flooding, and prevention of fatalities from LNG accidents,[26] are not reducible merely to the purpose of saving the most lives for the least money. Officials implementing safety programs in a given canyon, for example, cannot merely pursue risk reduction by prohibiting all building in the flood plain. Values such as the *freedom* of persons to buy their land and build on it, as well as rights to *privacy* and to *property*, have to be addressed alongside of, and perhaps instead of, the value of risk reduction.

Moreover, it seems clear that there is less justification for the government's telling a potential flood victim to what extent he can endanger himself on his land ('user' risk) than there is for the government's telling the LNG industry to what extent it can endanger the lives of other people ('third party' risk). As it is, however, Okrent and other proponents of the responsibility claim have masked the impact of their value judgments about matters such as freedom, property, and privacy. This masking has occurred, in large part, because of improper data comparisons (e.g., LGN versus flooded-canyon hazards) stemming from an oversimplification of the problem of safety allocation.

But if the purposes, values, and goals of the two programs cannot be reduced simply to issues of health or survival, then the opportunity costs of these programs are not reducible simply to excess deaths or ill health. Investing in risk reduction at liquefied natural gas (LNG) facilities, for example, may include one type of opportunity cost (e.g., the excess deaths caused by not spending the LNG monies, instead, to prevent canyon flooding), while investing in risk reduction to protect potential victims of canyon flooding may include another type of opportunity cost (e.g., loss of certain property rights dealing with one's ability to use his land as he wishes). Because the opportunity costs in the two programs are not the same (i.e., loss of life vs. loss of rights), it is misleading to charge merely that investors who do not follow the commensurability presupposition are 'responsible' for the excess deaths that could have been prevented through another investment. The charge is misleadingly accusative of the investor, both because *not* following the commensurability presupposition *also* would make one 'responsible' for other undesirable opportunity costs, e.g., loss of certain property rights, and because, if one grants that persons are always responsible for the opportunity costs of their actions, then there is never a way for any investor to avoid responsibility for some type of negative opportunity costs.

Thus, while an investor may be 'responsible' somehow for excess deaths, as Okrent claims, it is not obvious that he ought always to accept certain opportunity costs (e.g., violations of property rights) but always to avoid other opportunity costs (e.g., excess deaths caused by not maximizing the lives saved per dollar). This is because, as was explained in criticism of the rationality argument (Section 3.1.2), one is not bound always to maximize the lives saved per dollar. But if this is so, then one's alleged responsibility for excess deaths is not necessarily grounds for subscribing to the commensurability presupposition. Hence, even if one is somehow 'responsible' for excess deaths, this fact alone does not prove that one ought *not* to have taken the policy action allegedly resulting in these excess deaths. In other words, there is no clear connection between accepting the responsibility claim and therefore accepting the commensurability presupposition.

3.1.3.2. *The assumption that one is responsible for excess deaths.* More importantly, there may be little reason to accept the responsibility claim in the first place. This is because, in general, it is built on a concept of responsibility which is virtually unbounded. Its proponents assume that one is always responsible for all the opportunity costs (in this case, excess deaths) of his actions/investments/policy decisions. Their assumption is doubtful, however, because the causal connection between a particular investment and the opportunity cost of excess deaths is sometimes very tenuous. Let's see intuitively why this is the case, and then examine some more complex philosophical bases for questioning the causal connection between a particular investment and excess deaths.

Consider the case of a public official who decided to spend government funds to provide summer recreation programs to help keep disadvantaged young people safe and off the streets, rather than to help save local children who were victims of leukemia. Suppose also that it could be shown that more lives per dollar could be saved through the leukemia program, rather than through the recreation program. Following the responsibility claim, ought one to argue that the official was somehow responsible for the deaths of the children who otherwise might have been saved? Would the person funding the recreation programs be "killing people whose premature deaths could be prevented", as Okrent put it? Are the youths who use the summer recreation program likewise 'responsible' for killing leukemia victims? On all three counts, one might intuitively answer 'no or, at least, 'not necessarily'. This is likely because the causal connection between a particular funding decision and excess deaths is very tenuous.

In the case of Okrent's investor who pursues risk reduction for LNG facilities and thereby is 'responsible' for canyon deaths caused by failure to fund a canyon-flooding safety program, the causal connection between LNG investment and the 'excess' canyon deaths is tenuous because implementing the LNG program, *alone*, is not sufficient to cause the canyon program not to be funded and, therefore, is not sufficient to cause the excess deaths. Other people and other factors also contribute to the deaths, e.g., perhaps the victims themselves, if they are careless. But if persons besides the investor or decisionmaker are also responsible, then it is inaccurate simply to say that he is *as* responsible for the canyon deaths as the polluter is for deleterious health effects which his emissions are alone sufficient to cause. Clearly, where responsibility is divided among many persons/groups, each person (e.g., the decisionmaker) bears less than total responsibility.

Bayles argues against similar grounds for holding decisionmakers in developed countries responsible for lives lost in underdeveloped countries, and his remarks appear applicable here. He maintains that it is wrong to claim that people in developed nations are solely responsible for deaths in third-world countries, because such a thesis rests on "role responsibility, that of being ultimately in charge because of one's position". Although good reasons may exist for holding a person liable for another's death (e.g., holding the auto manufacturer liable in a car accident), says Bayles, the person allegedly liable cannot be said to have caused the death. This is because "role responsibility" is not, of itself, sufficient to cause a thing to occur. Individual stockholders of corporations, and individual citizens of countries, affirms Bayles, cannot by themselves effect a change in the policy of their corporations or their countries.[27] Likewise, to the extent that an individual policymaker does not by himself cause excess deaths, to that degree is he less responsible for them. Hence, even if proponents of the responsibility claim are correct, policymakers may be responsible only in a limited sense.

Third, the fact that failure to fund a given safety program is an act of omission, rather than an act of commission, suggests that responsibility for excess deaths (somehow partially attributable to this failure) may be mitigated. There are several reasons why it appears that one is more responsible for acts of commission than for acts of omission. Acts of commission are usually accomplished by *actively doing* something, while acts of omission are often the product of no active doing, intention, or choice: they may result simply from passivity. Also, since acts of commission require one to do something, rather than merely to accept what has been done, they are generally accomplished with more deliberation and purpose. Acts of omission often are characterized by little deliberation or purpose. Frequently, therefore, one is more responsible for acts of commission because they are more likely to be actively, deliberately, and purposively chosen.[28] But if one is often *more* responsible for acts of commission, and if failing to fund a particular safety program is an act of omission, not commission, then contrary to proponents of the responsibility claim, certain investors may not be fully responsible for 'killing people'.

Fourth, contrary to the responsibility claim, an investor may not be culpable for 'killing people' if he has no obligation to help save the greatest number of lives per dollar. One of the reasons why responsibility for the alleged opportunity costs of an action often exists is that one has an obligation not to violate the rights of others. One's responsibility, for example, for the opportunity costs of using the air as a receptacle for dangerous pollutants, exists in part because of an obligation not to violate other persons' rights to property and to equal protection. Were there no such rights, then it would be useless to blame someone for 'killing others'. In the case of one's alleged responsibility for excess lives lost, because of investing in a given safety program rather than in another, obligations appear to play little, if any, role. If such a responsibility exists, it is because of an *ideal* to enhance societal welfare, not because of an *obligation* to recognize rights.

Although it might be argued, following W. D. Ross, that one does have an obligation to benevolence, there are at least two reasons why there is no obligation to maximize the number of lives saved per dollar. First, such an obligation to maximization would be impossible to meet under all circumstances. *Ought* implies *can. Second*, one has no *right* to have society maximize his health and safety by means of the commensurability presupposition, but only a *right* not to be harmed unjustly by another.[29] This point is most obvious in cases where improvements in health and safety can be obtained most cost-effectively by preventing people from harming themselves. For example, if prohibition of alcohol were the most cost-effective way to save lives per dollar, it is not clear that citizens would have the *right* to prohibition programs. At best, if at all, one seems able to argue only that such programs are defensible on grounds that they help one to pursue the ideal of enhancing societal health and safety. But if maximization of health and safety is something to which one need have no *right*, then it is not clear that an investor is culpably responsible for not maximizing health and safety.

A related point is that assertions of responsibility for doing or not doing something are dependent for their meaningfulness on there being an "ongoing rule-regulated system of expectations" to which an action or inaction comes as an exception.[30] Applied to the responsibility claim, this means that one is culpable for not saving the greatest number of lives per dollar only if his failure to do so is an exception to a recognized, "ongoing rule-regulated system of expectations". Since there is no recognized rule to maximize only economic efficiency in saving lives, then there is no clear culpability for failure to do so.

Moreover, to say that one is responsible for lives lost because of funding a given safety program, just as one is responsible for the opportunity or social costs of his actions, is to presuppose a particular view of the state. Only if one believes that the state exists, not only to prevent harms, but also to increase/redistribute welfare, could one argue that decisionmakers are equally responsible both for recognizing rights to equal protection and for enhancing welfare. If one does not accept such an extensive role for the state, then it makes no sense to hold decisionmakers responsible for failures to increase welfare by saving the greatest number of lives per dollar. In either case, one cannot claim to have a *right* to have the state increase welfare by maximizing health and safety. And if not, then there are limited grounds for agreeing with the responsibility claim. This, in turn, means that appeal to the responsibility claim to support the commensurability presupposition is successful only if supported by good reasons relevant to a particular case. Providing these good reasons would presuppose, for example, (1) that the situation were one in which health/safety per dollar ought to be maximized; (2) that there were a clear, defensible causal connection between a particular investment and 'excess deaths'; (3) that the failure to fund a given program (in which excess deaths occurred) was not merely an indeliberate omission; and (4) that failure to prevent these excess deaths was somehow a violation of an obligation or of some right. This brings us to the question of how to judge whether the equity claim, the rationality claim, and the responsibility claim ought to apply to a given situation and what might constitute good reasons for appealing to these three claims. In other words, when might there be good reasons for appealing to one of these claims in order to support the commensurability presupposition?

3.2. *Factual Criteria for Using the Commensurability Presupposition: Simple Cases and Difficult Cases*

As Okrent's LNG/flooded canyon example suggests, when assessors naively make appeals for equity, rationality, and responsibility in evaluating the cost per life saved across opportunities, their arguments often fail. This is because they neglect to distinguish the simple cases, in which the appeals to consistency or equity work, from the difficult cases, in which they do not. As a consequence, they ignore the fact that even in choosing safety programs, society must promote many values in addition to safety.

Let us look at an example in which the appeals to equity and rationality are successful, and thereby discover why similar appeals are frequently not successful. Consider the case of auto safety. On the basis of cost per traffic fatality forestalled, one could easily argue that it probably makes more sense to increase monies for building medians on all major highways, than to step up funding for driver-education programs.[31] In this example, there appear to be at least three, good, factual reasons why the appeals to equity and rationality, across opportunities, are sensible, and why they have been used quite successfully in auto assessments by the U.S. Office of Technology Assessment.

First, there is a single *constituency*, the taxpayers, who bear the cost of the two alternative programs and a single constituency, automobile drivers and pedestrians, who receive the benefits of both programs. Moreover, in developed countries, the class of all automobile drivers and pedestrians very nearly approximates the class of all taxpayers. This means that the class of those receiving the benefits of both safety programs is almost the same as the class of those bearing the costs of both programs. As a consequence, it is much less likely that troubling questions of distributive equity will arise in this example, as compared to cases whose constituencies are vastly different.

Second, both programs (driver education and road construction) share a quite narrow *purpose* and *value*; improving the health, safety, and efficiency of automobile travel. On the contrary, consider what might happen if one were to compare two safety programs with quite different purposes and values. Suppose one were directed at the private sector, e.g., reducing fatalities from fires at private residences, while the others were directed at the public sector, e.g., reducing fatalities from fires in rental units or hotels. If these two cases were compared, then one would have to take different values into consideration. These might include the rights to private property and the responsibility to protect the common good. But, as a consequence of having to consider somewhat diverse goals and values for each safety program, it would be far less plausible merely to compare the two programs solely on the basis of their marginal cost per life saved.

Third, another reason for the success of the appeals to equity and rationality in the two auto safety cases is that the two programs address the same sorts of *hazards* having the same types of *effects*, i.e., the risks and benefits are in the same class. Obviously it is less question-begging to assume that two programs ought to be consistent, with respect to the value they place on saving life, if the effects of the two cases are similar. For example, it appears somewhat reasonable to compare two broad-spectrum chemical pesticides, each with similar effects, solely on the grounds of the commensurability presupposiiton, in order to determine how one might save the most lives for the least money. The comparison would be less reasonable, however, if the policy alternatives included both biological and chemical forms of pest control. This is because the options obviously have quite different costs, benefits, and effects, i.e., quite diverse ecological, agricultural, economic, medical, and political effects. In this latter case, the goals, values, and constituencies, as well as the effects, of the various pest reduction programs, are much more diverse than in the first pesticide case. For this reason the latter situation is much less amenable to formulations considering only economic efficiency in saving lives.

All this suggests that the 'hard cases', the safety-program comparisons in which a simple appeal to consistency (in marginal cost per life saved across opportunities) is unlikely to work, are those having (1) diverse constituencies; (2) different purposes and goals, and (3) risks, benefits, and effects of many different types. Likewise, use of the commensurability presupposition appears most likely to be reasonable in situations in which the constituencies, goals, risks, benefits, and consequences of the safety program are similar.

If Margaret Mead is right, then the difficult cases (in which use of the presupposition is not reasonable) very likely outnumber the simple ones in risk assessment. As a consequence, we may well wish to draw some problematic distinctions about how best to allocate funds. We may wish, for example, to spend more to save persons from slow, painful deaths than to save them from relatively quick, painless ones.[32] Or, we might wish to go to extraordinary lengths to save people who are suffering in extraordinary circumstances.[33] Moreover, as several authors have pointed out,[34] we may wish to spend more to save persons from uncompensated risks than from compensated ones. If so, then we may wish to spend more money in programs to avoid slow and painful deaths, or uncompensated risks, than in programs to avoid relatively quick and painless deaths, or compensated risks. Presumably, one might justify such marginal-cost differences in saving lives by arguing that the risks addressed by the two safety programs are quite different, and that greater monies could well be spent to avoid the more horrible sorts of deaths. In general, then, evaluating the similarity among the constituencies, goals, risks, benefits, and consequences of alternative safety programs provides some preliminary *factual* grounds for deciding how to spend funds for risk abatement.

In general, this preliminary observation is analogous to another observation which appears to be quite obvious. Just as unequal treatment tends to be less justifiable, in a given situation, to the degree that all persons are equal in all relevant respects, so also unequal treatment of potential victims, across risks, appears to be less justifiable to the degree that the constituencies, goals, risks, benefits, and consequences of the safety programs are similar. Consistent with the principle that equal beings in similar situations ought to be treated equally, this observation (about constituencies, etc.) specifies relevant respects in which safety situations might be similar or dissimilar. As such, this observation reveals the *factual* conditions under which discrimination is likely to be justified or not. What might be some *ethical* conditions under which discrimination is likely to be justified among safety programs?

3.3. *Ethical Criteria for Using the Commensurability Presupposition*

Inasmuch as decisions about using the commensurability presupposition can be addressed by ethical criteria, these tend to focus on the issue of equal protection. If, following the earlier discussion in Section 3.1, use of this presupposition cannot be justified by a simple, *general* appeal to equal protection, then it becomes important to know what constitutes good reasons, in a *particular* case, for following the commensurability presupposition. The argument of the forthcoming pages is that two principles furnish some useful guidelines regarding this presupposition. I call them, respectively, (1) the principle of *prima-facie* egalitarianism and (2) the principle of everyone's advantage.

3.3.1. *The Principle of Prima-Facie Egalitarianism*

According to the principle of *prima-facie* egalitarianism, although equality (sameness) of protection is not desirable in all situations (for some of the reasons spelled out in Section 3.1), it ought to be adopted as a *prima-facie* principle. This means that it is presumed applicable in a particular case unless it is showed to be inapplicable. According to the principle of *prima-facie* egalitarianism, only unequal protection (spending unequal amounts, per life saved, in order to reduce risk) requires justification.[35]

With the proponents of the equity claim, advocates of this principle believe that equal protection is desirable. Unlike them, however, they believe that good reasons can sometimes be shown to argue against spending the same amounts, per life saved, in all risk situations. This means that, while the principle of *prima-facie* egalitarianism is not absolute and does not hold for all cases, it is *prima facie* desirable. Therefore, the burden of proof ought to be placed on the person who wants to 'discriminate' through unequal protection. Not to do so would be to encourage expenditures for risk reduction to become the political footballs for various interest groups. For example, it would be to encourage the U.S. to continue to spend millions of dollars to save the lives of middle-aged, overweight heart-attack victims, but virtually nothing to save the lives of 4-month old victims of sudden infant death syndrome. Whether such expenditures are justifiable or not, the proponents of the principle of *prima-facie* egalitarianism believe that the presupposition ought to be in favor of equal protection, and for at least four reasons:

(1) the comparison class is all humans, and all humans have the same capacity for a happy life,[36]
(2) free, informed, rational people would agree to the principle;[37]
(3) it provides the basic justification for other important concepts of ethics; it is a presupposition of all schemes involving justice, fairness, rights, and autonomy;[38] and
(4) equality of treatment is presupposed by the idea of law; "law itself embodies an ideal of equal treatment for persons similarly situated."[39]

As was already pointed out (in Section 3.2), one might have good reasons for spending unequal amounts to save lives in alternative safety programs if those programs have different constituencies, goals, risks, benefits, and consequences. Hence, according to the principle of *prima-facie* egalitarianism, the fact that a law, action, or policy discriminates among persons does not necessarily make it wrong, contrary to what proponents of the commensurability presupposition believe. Discrimination among safety programs is wrong only if it is done arbitrarily or for irrelevant reasons.[40]

3.3.2. *The Principle of Everyone's Advantage*

Although a clear and precise line between relevant and irrelevant reasons for discrimination (with respect to equal protection) is not evident in every case, at least one ethical criterion for drawing such a line comes to mind. This is what I call the 'principle of everyone's advantage', the thesis that unequal protection (spending different amounts, per life saved, across risk opportunities) among persons is justified if the 'discrimination' works to the advantage of everyone.[41] (The principle is not equivalent to the Pareto criterion, since 'advantage' is not employed in it in a purely economic sense.[42])

For example, suppose government regulations require lives to be saved at a cost of $\$x$ each in most safety programs. But suppose also that if lives were saved at a cost of $\$2x$ in a particular LNG (liquefied natural gas) program, then as a consequence everyone would be better off, in economic as well as noneconomic terms, because of the increased safety of the facility. But if everyone would be better off, then the discrimination would be justifiable. Of course, the problematic aspects of this example are judging whether everyone indeed would be better off as a consequence, and defining what it is to be better off. In this case, the judgment that the discrimination in fact will work for the advantage of everyone might be a function of several claims. These include, for example, the 'fact' that a $\$2x$ cost per life saved might decrease opposition of local residents to the LNG facility. Or, it might cause more equity in the distribution of goods and resources than otherwise would have occurred without the initial discrimination in favor of persons put at risk by the LNG facility.

Admittedly, drawing the conclusion that it would be to everyone's advantage to discriminate in this manner rests on a chain of tenuous causal inferences and assumptions, some of which arise out of ethical theory. In spite of the obvious practical difficulties in judging whether a discrimination in fact will contribute to everyone's advantage, this principle appears to be the most promising theoretical candidate for an ethical criterion to determine acceptable discrimination. This is because almost any other principle would be open to the charge that it sanctioned using some persons as means to the ends of others.[43]

Because humans may not be used as *means* to some end (a principle which we take to be self-evident), fulfilling the principle of everyone's advantage is a *necessary* condition for justifying discrimination among potential victims protected by alternative safety programs.[44] It is also a *sufficient* condition, since presumably any legitimate grounds for opposing discrimination (e.g., the existence of certain rights) would be equivalent to the claim that the discrimination did not serve the authentic advantage of everyone.

One major objection to this principle is that, as Dr. Douglas MacLean of the University of Maryland puts it, no technology-related 'discrimination' against equal (the same level of) protection will ever, or has ever, worked to the advantage of everyone. On the contrary, it is not evident to me that no such discrimination has met, or will ever meet, this criterion. To determine whether a given discrimination might work to everyone's advantage, it seems that one would need to employ some fairly sophisticated economic and ethical analyses (see Sections 3.3.2 and 4 of this essay). Moreover, even if MacLean is correct, his point appears not to be a damning one against use of the principle. This is because, in at least some cases, all victims could likely be adequately compensated, if a given discrimination were not to their advantage.[45]

4. FUTURE DIRECTIONS AND THE COMMENSURABILITY PRESUPPOSITION

The principle of everyone's advantage provides both a necessary and a sufficient condition for justifying discrimination. This realization outlines the theoretical constraints governing use of the commensurability presupposition. That is, the commensurability presupposition (which requires sameness of marginal costs across opportunities) ought not to be held, and discrimination among potential victims in alternative safety programs is justifiable, provided that the discrimination works to everyone's advantage. Given this insight, the task facing risk assessors is three-fold, if they wish to apply these results to actual decisions about allotments of funds and to avoid uncritical acceptance of analytic risk methodology. Each of these tasks requires substantial ethical analysis, in order to determine whether, in a particular situation, withholding use of the commensurability presupposition will indeed work to everyone's advantage.

What are the three tasks facing future researchers? First, following Section 3.1.1, assessors must ascertain, in a given safety expenditure decision, whether everyone's interests can be accorded the same concern and respect, even if the commensurability presupposition is not held. In other words, one necessary condition for a given discrimination's serving everyone's advantage is that, as Dworkin says, it not violate anyone's right to equal concern or respect. *Second*, following Sections 3.1.1 and 3.3.1, and considering the morally relevant reasons justifying unequal treatment (as a reward for merit, etc.), assessors must determine, in a particular safety-expenditure decision, whether recognition of any of these (or other) morally relevant reasons, in this specific case, works to everyone's advantage. If decisionmakers do not follow the commensurability presupposition, it is necessary that this be shown. *Third*, following Section 3.2, assessors must ascertain, in a given safety-expenditure decision, whether there are morally relevant *factual* dissimilarities among the constituencies, goals, and consequences of alternative safety programs, such that these dissimilarities justify failure to subscribe to the commensurability presupposition. Although determining these facts is not necessary for justifying discrimination, the absence of factual dissimilarities could constitute a strong argument for using the commensurability presupposition in a given case.

In addition to specific research consisting of ethical analysis of each of these three issues, risk assessors are also faced with some complex economic and social-scientific work, if they are to discover the limits of the use of the commensurability presupposition.

As I pointed out in Section 3.3.2, one of the main tasks in this regard is establishing a network of plausible causal inferences enabling one to determine, given safety expenditures not consistent with the commensurability presupposition, whether these expenditures indeed are likely to work to everyone's advantage.

As is already clear from this overview of the tasks still to be accomplished, this chapter met several *theoretical* aims. It has exposed the limitations of wholesale use of the commensurability presupposition, suggested factual and ethical conditions under which the presupposition ought and ought not to be used, and provided a necessary and sufficient ethical condition for discriminating among potential victims in safety programs. What remains is to *apply* these ethical and methodological conclusions to specific risk-analysis cases. This application will require both philosophical (ethical and methodological) and scientific (especially economic) expertise.

5. SUMMARY AND CONCLUSIONS

If the preceding discussion is correct, then there are strong grounds for rejecting absolute acceptance of the commensurability presupposition and the reasons typically used to support it (the equity claim, the rationality claim, and the responsibility claim). Rather, situations of costing lives, across opportunities, appear to fall into two classes, one of which is more amenable to use of this presupposition, and one of which is less so. One set of risk-abatement situations is generally more amenable to use of this presupposition because the safety programs at issue have similar constituencies, goals, risks, benefits, and consequences. The other set is less amenable to it because the programs being compared have dissimilar constituencies, goals, risks, benefits, and consequences.

In terms of *ethical* criteria for using the commensurability presupposition in given safety situations, I argued that two principles are useful. The principle of *prima-facie egalitarianism* establishes the *prima facie* desirability of giving the same protection to prospective victims whose needs are addressed by various safety programs; further, it places the burden of proof on the person wishing not to give equal (the same level of) protection. The principle of *everyone's advantage* provides a necessary and sufficient condition for justifying unequal protection in a given situation.

All these arguments suggest that there are strong reasons for believing that the marginal costs of saving lives need not always be the same, across opportunities, and that there is at least one ethical principle according to which people may receive unequal 'protection' with respect to this marginal cost. Employment of this principle requires, however, extensive ethical analysis and evaluation of numerous causal inferences, in order to determine what courses of action, in fact, will contribute to the goodness of people's lives or to their advantage. This means that the whole issue of the marginal cost of saving lives, across opportunities, is not so much a matter of *economic consistency* as of *ethical analysis*.

NOTES

1 See, for example, L. Lave and E. Seskin, 'Air Pollution and Human Health', *Science* **169** (3947), (1970), 723–733; hereafter cited as: APHH. See also D. Rice, *Estimating the Cost of Illness*. PHS Publication No. 947–6, U.S. Government Printing Office, Washington, D.C., 1966.

2 In two essays, I argue in favor of analytic assessment techniques. See 'Technology Assessment and the Problem of Quantification', in *Philosophy and Technology*, eds. R. Cohen and P. Durbin, Boston Studies in the Philosophy of Science, D. Reidel, Boston, 1983, forthcoming, and 'Die Technikbewertung und das Problem ihrer genauen Berechnung', in *Technikphilosophie in der Diskussion*, ed. F. Rapp, Vieweg Verlag, Weisbaden, 1982, pp. 123–138.

[3] See A. Kneese, S. Ben-David, and W. Schulze, 'A Study of the Ethical Foundations of Benefit–Cost Analysis Techniques', Working paper, 1979, pp. 23 ff.; hereafter cited as: Foundations.

[4] This method consists of experts' formalizations of past societal policy regarding various risks. Followed by assessors such as Starr and Whipple, the technique rests upon the assumption that past behavior regarding risks, benefits, and their costs is a valid indicator of present preferences. In other words, the "best" risk–benefit trade-offs are defined in terms of what has been "traditionally acceptable," not in terms of some other (e.g., more recent) ethical or logical justification. This, of course, involves the assumption that past behavior is normative, whether it was good or bad, or right or wrong. For this reason, some theorists have argued that the method of "revealed preferences" is too conservative in making consistency with past behavior a sufficient condition for the correctness of current risk policy (see B. Fischhoff, *et al.*, 'How Safe Is Safe Enough?', *Policy Sciences* **9** (2), (1978), 149; hereafter cited as: Safe). See also note 5 below and Chapter Two of this volume.

[5] Unlike the method of "revealed preferences", that of "expressed preferences" does not rely on past policy. Developed by assessors such as Fischhoff and Slovic, this approach consists of using questionnaires to measure the public's attitudes toward risks and benefits from various activities. The weakness of this method, of course, is that often what people say about their attitudes toward various risks appears inconsistent with how they behave toward them. Some theorists also view the method as too variable since it takes no account of past societal behavior but only relies on selected responses as to what people say they believe about risks (see Fischhoff *et al.*, Safe, p. 149). See also C. Starr, *Current Issues in Energy*, Pergamon, New York, 1979, p. 7; hereafter cited as *Energy*. See also Chapter Two of this volume.

[6] See C. Starr and C. Whipple, 'Risks of Risk Decisions', *Science* **208** (4448), (1980), 1118; hereafter cited as: Risks, and J. Hushon, 'Plenary Session Report, in the Mitre Corporation, *Symposium/Workshop on Nuclear and Nonnuclear Energy Systems: Risk Assessment and Governmental Decision Making*, The Mitre Corporation, McLean, Virginia, 1979, p. 748; hereafter cited as: Hushon, Report, and Mitre, *Risk*. See also D. Okrent, 'Comment on Societal Risk', *Science* **208** (4442), (1980), 374; hereafter cited as: Risk, and M. Maxey, 'Managing Low-Level Radioactive Wastes', in J. Watson (ed.), *Low-Level Radioactive Waste Management*, Health Physics Society, Williamsburg, Virginia, 1979, p. 401; hereafter cited as: Maxey, Wastes, and Watson, *Waste*. Finally, see J. D. Graham, 'Some Explanations for Disparities in Lifesaving Investments', *Policy Studies Review* **1** (4), (May 1982), 692–704, as well as C. Comar, 'Risk: A Pragmatic De Minimus Approach', *Science*

203 (4378), (1979), 319; hereafter cited as: Pragmatic, and B. Cohen and I. Lee, 'A Catalog of Risks', *Health Physics* **36** (6), (1979), 707; hereafter cited as: Risks.

7 F. Hapgood, 'Risk–Benefit Analysis', *The Atlantic* **243** (1), (January 1979), 28; hereafter cited as: RBA. See also J. Norsigian, in Congress of the U.S., *Fertility and Contraception in America.* Hearings before the the Select Committee on Population, 95th Congress, Second Session, III, No. 4, U.S. Government Printing Office, Washington, D.C., 1978, p. 375. Norsigian points out that investments in contraceptive research and development are inequitable because the cost per male life saved is much greater than the cost per female life saved.

8 See note 6.

9 Starr and Whipple, Risks, p. 1118 (note 6).

10 Hushon, Report, p. 748 (note 6).

11 W. Häfele, 'Energy', in C. Starr and P. Ritterbush (eds.), *Science, Technology, and the Human Prospect*, Pergamon, New York, 1979, p. 139; hereafter cited as: Häfele, Energy, and Starr and Ritterbush, ST. See also A. Lovins, 'Cost–Risk–Benefit Assessment in Energy Policy', *George Washington Law Review* **45** (5), (1977), 941; hereafter cited as: CRBA.

12 Okrent, Risk, p. 373 (note 6).

13 B. Fischhoff, *et al.*, 'Which Risks Are Acceptable?', *Environment* **21** (4), (May 1979), 17; hereafter cited as: Risks.

14 Okrent, Risk, p. 374 (note 6). See also Starr, Energy, pp. 22–23, and D. Okrent and C. Whipple, *Approach to Societal Risk Acceptance Criteria and Risk Management*, PB-271 264, U.S. Department of Commerce, Washington, D.C., 1977, pp. 3–11; hereafter cited as: *Approach.* Others who share this point of view include: C. Sinclair, *et al., Innovation and Human Risk*, Centre for the Study of Industrial Innovation, London, 1972, pp. 11–13, and Committee on Public Engineering Policy, *Perspectives on Benefit–Risk Decision Making*, National Academy of Engineering, Washington, D.C., 1972, p. 12. These latter two works are hereafter cited (respectively) as: Sinclair, *Risk*, and Committee, *Perspectives.*

15 Okrent, Risk, p. 375 (note 6). See Starr, *Energy*, p. 10 (note 5), and L. Sagan, 'Public Health Aspects of Energy Systems', in H. Ashley, *et al.* (eds.), *Energy and the Environment*, Pergamon, New York, 1976, p. 89; hereafter cited as Sagan, Public, and Ashley, *Energy.*

16 R. Dworkin, *Taking Rights Seriously*, Harvard University, Cambridge, 1977, p. 273; hereafter cited as: *Rights.*

17 Dworkin, *Rights*, pp. 267–279.

18 Dworkin, *Rights*, pp. 267–279.

19 T. C. Bergstrom, 'Living Dangerously', in D. Okrent (ed.), *Risk–Benefit Methodology and Application*, UCLA School of Engineering and Applied Science, Los Angeles, 1975, p. 233.

20 For a similar argument, see M. Bayles, *Morality and Population Policy*, University of Alabama Press, University, Alabama, 1980, pp. 28–31; hereafter cited as: *Morality*.

21 G. L. S. Shackle, *Epistemics and Economics: A Critique of Economic Doctrines*, Cambridge University Press, Cambridge, 1972, p. 82; hereafter cited as *EE*.

22 Shackle, *EE*, p. 82.

23 E. Mishan, *Cost–Benefit Analysis*, Praeger, New York, 1976, pp. 153–174; hereafter cited as: *CBA*. See also Peter Self, *Econocrats and the Policy Process*, Macmillan, London, 1975, p. 68.

24 Okrent, Risk (note 6), pp. 372–375.

25 H. Siebert, *Economics of the Environment*, Lexington Books, Lexington, Massachusetts, 1981, pp. 16–17; hereafter cited as: Siebert, *EE*.

26 Okrent, Risk (note 6), pp. 372–375.

27 M. Bayles, *Morality* (note 20), p. 121.

28 Alan Gewirth, *Reason and Morality*, University of Chicago Press, Chicago, 1978, pp. 222–240, argues that many omissions are morally reprehensible, and I do not wish to take issue with (what I believe is) an essentially correct point. My thesis is that one is likely more responsible for acts of commission than for acts of omission. Gewirth's book is hereafter cited as: *Reason*.

29 Gewirth, *Reason*, p. 226.

30 This point is also made by Gewirth, *Reason*, p. 223.

31 Congress of the U.S., Office of Technology Assessment, *Technology Assessment of Changes in the Future Use and Characteristics of the Automobile Transportation System: Summary and Findings*, 2 vols, U.S. Government Office, Washington, D.C., 1979, II, pp. 207–208, 219.

32 Cited by Maxey, Wastes (note 6), p. 401.

33 E. Lawless, *Technology and Social Shock*, Rutgers University, New Brunswick, N.J., 1977, pp. 509–512.

34 Kneese, *et al.*, Foundations (note 3), p. 26. K. Shrader-Frechette, *Nuclear Power and Public Policy*, D. Reidel, Boston, 1980, pp. 108 ff.

35 For an excellent defense of this position, see W. K. Frankena, 'Some Beliefs about Justice', in J. Feinberg and H. Gross, *Philosophy of Law*, Dickenson, Encino, California, 1975, pp. 252–257; hereafter cited as: Frankena, Beliefs, in Feinberg and Gross, *POL*. See also W. K. Frankena, *Ethics*, Prentice-Hall, Englewood Cliffs, N.J., 1963, p. 41. '*Prima facie* egalitarians' (Frankena calls them 'procedural egalitarians') are to be distinguished from substantive egalitarians, who believe that there is some factual respect in which all human beings are equal. *Prima facie* egalitarians deny that there is some such factual respect. I am grateful to Dr. Douglas MacLean of the University of Maryland for suggesting that I use the term '*prima facie* egalitarian'.

36 W. T. Blackstone, 'On Meaning and Justification of the Equality Principle', in Blackstone, *Equality*.

37 See note 36. John Rawls, 'Justice as Fairness', in Feinberg and Gross, *POL* (note 35), p. 284, also makes this point; hereafter cited as Rawls, Fairness.

38 For arguments to this effect, see M. C. Beardsley, 'Equality and Obedience to Law', in Sidney Hook (ed.), *Law and Philosophy*, New York University Press, New York, 1964, pp. 35–36; hereafter cited as: Equality. See also Isaiah Berlin, 'Equality', in Blackstone, *Equality* (note 36), p. 33; Frankena, Beliefs (note 35), pp. 250–251; M. Markovic, 'The Relationship Between Equality and Local Autonomy', in W. Feinberg (ed.), *Equality and Social Policy*, University of Illinois Press, Urbana, 1978, p. 93; hereafter cited as Markovic, Relationship, and Feinberg, *Equality*. See also Rawls, Fairness (note 35), pp. 277, 280, 282, and G. Vlastos, 'Justice and Equality', in R. B. Brandt (ed.), *Social Justice*, Prentice-Hall, Englewood Cliffs, N.J., 1962, pp. 50, 56; hereafter cited as: Vlastos, Justice, and Brandt, *Justice*.

39 J. R. Pennock, 'Introduction', in J. R. Pennock and J. W. Chapman (eds.), *The Limits of Law*, Nomos XV, The Yearbook of the American Society for Political and Legal Philosophy, Lieber-Atherton, New York, 1974, pp. 2, 6; hereafter cited as: Pennock and Chapman, *LL*.

40 R. A. Wasserstrom 'Equity' in Feinberg and Gross, *POL* (note 37), p. 246, also makes this point. Even the Fourteenth Amendment, under the equal protection clause, does not prohibit all discrimination, but merely whatever is "arbitrary". In this regard, see N. Dorsen, 'A Lawyer's Look at Egalitarianism and Equality', in J. R. Pennock and J. W. Chapman (eds.), *Equality*, Nomos IX, Yearbook of the American Society for Political and Legal Philosophy, Atherton Press, New York, 1967, p. 33; hereafter cited as: Look in *Equality*.

41 See John Rawls, *A Theory of Justice*, Harvard University Press, Cambridge, 1971; hereafter cited as: Rawls, *Justice*. See also Charles Fried, *Right and Wrong*, Harvard University Press, Cambridge, 1978; and Alan Donagan, *The Theory of Morality*, University of Chicago Press, Chicago, 1977. See also S. I. Benn, 'Egalitarianism and the Equal Consideration of Interests', in Pennock and Chapman, *Equality* (note 50), pp. 75–76. See also Frankena, *Ethics* (note 37), pp. 41–42.

42 I am grateful to Dr. Toby Page of California Institute of Technology, for pointing out the question of whether the Principle of Everyone's Advantage is identical to the Potential Pareto criterion. There appear to be two reasons why they are not the same. *First*, the principle requires that everyone's advantage be served *in fact*, and that compensations be carried out, if everyone's advantage requires it. The Pareto criterion, however, does not require that the compensations actually be carried out. *Second*, the principle defines

"advantage" as overall welfare (including noneconomic well-being), whereas the Pareto criterion defines "advantage" in a purely economic sense. As was pointed out earlier, in Section 3.1.1 of this chapter, serving everyone's advantage might include according them their rights to equal concern and respect. Such rights, however, do not fall within the scope of the Pareto definitions of 'advantage'.

43 W. K. Frankena, 'The Concept of Social Justice', in R. B. Brandt (ed.), *Social Justice*, Prentice-Hall, Englewood Cliffs, N.J., 1962, pp. 10, 14.

44 In Section 4 of this chapter, we observed that discrimination in safety programs might be less justifiable to the degree that the programs shared similar constituencies, goals, risks, benefits, and consequences. Interpreting this preliminary observation in the light of the principle of everyone's advantage, we can now affirm that discrimination (among potential victims affected by alternative safety pograms) is *likely* to be justifiable to the degree that the programs have dissimilar constituencies, goals, risks, benefits, and consequences – *provided that* the discrimination works to the advantage of everyone.

45 Continuing our private conversation on the issue of whether any discrimination might work to everyone's advantage, Dr. MacLean's response to my rejoinder about compensation is that it would be "practically impossible" to accomplish such compensation. To this response, at least two points can be made. *First*, unless compensation is attempted in the cases in which it appears reasonable and equitable to try it, one will never know if it might be successful. Hence the only way not to beg the question, of whether compensation might work, is to try it. *Second*, the move toward compensation is at least plausible since welfare economists such as Mishan have discussed recognition of amenity rights, which would likely involve at least some cases of compensation. Recognition of amenity rights might require governments to examine the whole set of social costs (including imposed risks and various discriminations) in our society. Compensating the victims of such costs need not be more complex, in principle, than providing for the many current types of income-tax deductions and government subsidies prevalent today.

CHAPTER FOUR

OCCUPATIONAL RISK AND THE THEORY OF THE COMPENSATING WAGE DIFFERENTIAL

1. INTRODUCTION

Industrial risks and injuries are one way of marking the differences in privilege and status between employers and employees. They signify a fundamental clash of interests that goes to the heart of social inequality. Noting that workplace and industrial casualties are statistically at least three times more serious than street crime,[1] some policy experts argue that there should be no double standard for occupational and public exposure to various gases, chemicals, particulates, radiation, noise, and other forms of environmental pollution; they believe that industrial employees should be protected as much as the public is. According to proponents of this point of view, workers ought not to have to trade their health and well-being for wages. Moreover, they maintain, paying a person to put himself at risk at work is not substantially different from murder for hire. Sanctioning this belief, Judge Patrice de Charette, a French magistrate, caused substantial controversy when, in 1975, he and a deputy went to a refinery to arrest and imprison the plant manager where a worker had been killed in an industrial accident. When he was denounced by higher French authorities, de Charette maintained: "I don't see why it is less serious to let men die at work than it is to steal a car."[2]

Those who agree with the double standard for worker and public exposure to risk usually maintain that the additional wages received by workers in hazardous occupations compensate them for their risks. They also claim that occupational risk is overemphasized and sensationalized by the "danger establishment"[3], and that most countries, notably the U.S., have unacceptable 'rigid standards' for workplace risks. For example, a recurrent target of ridicule, for those who believe that U.S. occupational safety standards are too strict, is 'the portable toilet standard for cowboys' which was set by the U.S. Occupational Safety and Health Administration (OSHA).[4]

Heated though it is, disagreement over occupational risk is nothing new. Workplace risks and controversy over them originated at least as early as the emergence of a division of labor between manual and nonmanual labor. In fact, the Greek word for work, *ponos*, has the same root as the Latin word for sorrow, *poena*, which also means penalty. The French word, *travailler*, to work, is derived from a Latin word referring to a kind of torture. Ancient Greek and Roman writings are filled with references to the diseases peculiar to one or another profession. Later, during the Renaissance, miners and metal workers became the first subjects of medical research into diseases of the workplace. Perhaps the first publication to address occupational hazards and their prevention was a booklet written in Germany in 1472. It told goldsmiths how to avoid poisoning by mercury and lead. In 1556, in his treatise on the mining industry, the German mineralogist, Agricola, wrote the first known review of miners' health problems. He noted that some women who lived near the mines of the Carpathian Mountains in Eastern Europe had lost seven successive husbands to mine-related accidents and diseases. Besieging his medical colleagues and statesmen to make workplaces safer, the Italian physician, Ramazzini, wrote *Diseases of Workers* in 1700.[5]

Despite the historical knowledge that various risks and diseases are associated with particular jobs, surprisingly little has been done in a number of cases to avoid or to reduce substantially those known risks. As J. K. Wagoner of the U.S. National Institute for Occupational Safety and Health (NIOSH) observes, two centuries have passed since Percival Pott linked coal tars to the scrotum cancer that killed young chimney sweeps in England. Yet, he claims, "thousands of coke-oven workers in steel mills around the world continue to inhale the same deadly substances, and they are dying of lung cancer at ten times the rate of other steel workers."[6]

One reason for the continuing U.S. controversy over workplace hazards, and over whether to employ a double standard for public and occupational risk exposures, is that U.S. standards for health in the workplace appear to permit greater risks than do those of many other nations. In terms of permissible levels of chemicals in the work environment, for example, U.S. regulations are less strict than those of the Federal Republic of Germany, the German Democratic Republic, Sweden, Czechoslovakia, and the U.S.S.R. Standards in Argentina, Great Britain, Norway, and Peru are approximately the same as those in the United States.[7]

Since Soviet activities are frequently ignored in the U.S., it is a little-known fact that the U.S.S.R. has a long tradition of providing for worker health and safety. In 1923, the U.S.S.R. founded the first hospital devoted entirely to the study and treatment of occupational diseases. No such hospital exists in the U.S. Admittedly, relative enforcement patterns are not known and, although maximum-allowable-concentration (MAC) values may be lower in the U.S.S.R., control there is likely far less stringent than in western countries. If so, then despite safer standards in the U.S.S.R., risk there could be higher. Regardless of whose enforcement patterns are better, however, the U.S.-U.S.S.R. comparison raises a number of interesting philosophical questions. Among these are whether the Soviets have a more or less desirable risk philosophy than do their American counterparts. Another question is why Soviet MAC values are lower, "often by a factor of ten or more," than corresponding U.S. standards, even though the U.S.S.R. must confront many of the same problems that the U.S. faces (for example, asbestos production has soared, and the petrochemical industry is growing much faster than the rest of the economy).[8] Yet another question is whether one can ethically justify workplace MAC values, which are sometimes higher than corresponding values for public exposure, on the grounds that workplace exposure time is shorter than that for the public. Perhaps most important to the question of whether risk standards *ought to be* improved is whether lower MAC values are even technically possible. And, if they are possible, whether they would be so costly as to jeopardize the economic well-being and the technological progress which have resulted in enormous improvements in human welfare. Or, if they are possible, whether they would be so costly that most workers and citizens would not be willing to pay for them by raising the price of risk-related goods and services.

2. THE THEORY OF THE COMPENSATING WAGE DIFFERENTIAL

Although a variety of factors are likely responsible for the more lenient occupational safety standards in the U.S., as compared to those in other countries, at least one of the reasons for the disparity is less emphasis in the U.S. on equity. Typically, U.S. standards allow much higher exposure levels, to particular pollutants, for workers than for the public. In large part, this is because U.S. policymakers do not believe that equity requires occupational and public exposure levels to be the same. For example, the U.S. maximum permissible dose of whole-body radiation which can be received annually by the public is 500 millirems; the maximum permissible dose, for the same time period, for industrial workers is 5000 millirems, or ten times as much radiation.[9]

The main reason why equity is generally not thought to demand the same standard for occupational and public exposure to various pollutants is that the two types of exposures are not thought to be analogous. According to proponents of the method of revealed preferences (for evaluating risks),[10] for example, occupational risks are usually defined as *voluntary* risks while public risks are defined as *involuntary*; since involuntarily imposed risks ought to meet more stringent safety requirements, they maintain, the double standard for occupational and public risks is reasonable.[11] Also, they claim, risks accepted 'voluntarily', through one's occupation, can be regulated by means of standards less strict than those applied to public risks, precisely because people are compensated (through their wages) for the higher workplace risks that they bear. According to Chauncey Starr, one of the preeminent proponents of the method of revealed preferences, the risk entailed by a particular occupation is directly proportional to the cube of the wages for that occupation; as the risk increases, so do the wages.[12] Admittedly, however, the wage-risk relationship does not always appear to be so simple, especially in western countries. Many factors, in addition to risk, determine the wages received for given work. Some of these include the degree of education or training necessary for the job; the extent to which people are available, in a particular locale, to perform the work; or the degree of physical strength required to do the task. Hence, although there is some sort of wage-risk relationship, such that as job risks increase, so do wages, this relationship may not be nearly so simple, in all cases, as Starr supposes.

2.1. *Current Acceptance of the Theory of the Compensating Wage Differential*

Starr's view, widely accepted among risk assessors, especially among those who follow the method of revealed preferences, is part of the classic theory of the compensating wage differential. The fundamental economic principles of this theory were formulated long ago by Adam Smith. As Smith expressed it, "the whole of the advantages and disadvantages of the different employments of labor" continually tend toward equality because the wages vary according to the hardship of occupation; on this theory, men exposed to a risky workplace had advantages and disadvantages whose sum was equal to those to the men not exposed to such risks, because those in the high-risk occupations were provided with higher rates of pay than were those in low-risk jobs.[13]

According to proponents of the theory of the compensating wage differential, a double standard regarding worker and public risk is acceptable because those in high-risk jobs voluntarily agree to 'trade' some degree of workplace safety for higher wages. In other words, the classic solution to the problem of how to control occupational risks, and how to decide which worker risks are acceptable, is to use an 'economic fix', a market mechanism, for setting standards.[14]

2.2. *Arguments for the Theory of the Compensating Wage Differential*

In arguing for a market mechanism, the compensating wage differential, to resolve the problems of equity raised by the double standard for occupational and public risk, risk assessors, economists, and public policymakers generally employ at least four arguments. Let's examine each of them.

2.2.1. *The Welfare Argument*

One approach is to use a welfare-based argument. Its proponents maintain that "insistence on uniform hazard regulations will inevitably lead to . . . detrimental" results. They claim that this is because the double standard enables those in high-risk occupations to boost "their income status above what it would otherwise have been. If all jobs were required to be as safe as the most highly paid white-collar positions, the income status of those at the bottom of the income scale would be lowered further. Wage premiums for risk do exist, but they are not sufficient to offset all of the other factors generating the low-income status of the workers who receive them."[15] In other words, advocates of this argument maintain that the double standard for risk enhances the welfare of low-income groups because it provides them with higher wages than would a uniform standard. As Viscusi puts it, "if coke-oven workers are willing to endanger their lives in return for substantial salaries, or if India chooses to develop nuclear energy as the most promising energy source for its long-term development, government efforts to interfere with these decisions will reduce the welfare of those whose choices are regulated."[16]

Although the welfare argument is highly persuasive, in that it correctly emphasizes the importance of worker autonomy over government intervention, it is premised on a number of assumptions which are highly doubtful. Perhaps the most basic of these is that worker *preferences* are authentic indicators of desirable *values*, or at least that workers are better able to determine what is in their best interests than is government. However, in many cases, preferences are not legitimate indicators of authentic welfare, as can be seen if one examines some persons' preferences for particular marriage partners or for dangerous habits, such as smoking. It is well known to philosophers that preferences merely indicate demands, regardless of whether they are desirable demands or not, whereas welfare is concerned only with legitimate demands. Preferences reveal what people want. Their welfare, however, is determined by their having correct wants.[17]

Another questionable assumption of the welfare argument is that it is ethically acceptable to allow persons to trade their health and safety for money. Clearly, however, some such trade-offs would be wrong, e.g., those in which one allowed himself to be cruelly tortured in exchange for money. They might be wrong, either because they failed to acknowledge someone's rights, or because they did not respect the dignity of humans, or because they allowed the perpetrator (e.g., of the torture) to behave in reprehensible ways, or because they permitted one to use another human as a means to an end, when humans ought to be treated only as ends. In other words, it is not ethically acceptable, generally, to allow persons to trade their health and safety for money because person *A*'s consent is not a sufficient condition for the morality of person *B*'s actions affecting person *A*, even if *B* compensates *A* financially. Although they are often necessary conditions, consent and compensation are not sufficient conditions for the morality of an action, because the moral quality of an act is also determined by various rights, duties, and agreements. But if this is so, then it is not adequate to defend the theory of the compensating wage differential merely by appealing to notions of compensation, consent, or preferences.

2.2.2. *The Market-Efficiency Argument*

A second argument for accepting the theory of the compensating wage differential is that "market allocations of individuals to jobs will promote efficient matchups in many instances. If the worker bears all of the harm associated with the risk and if he is cognizant of his own particular risk, not simply the average risk for all, he will select his job optimally . . . workers are not in jobs at random and the market promotes the most efficient matchups."[18] For example, says Viscusi, "Blacks with the gene for sickle-cell anemia may incur a greater risk of harm from the low-oxygen conditions faced by a pilot, and female mail sorters have a greater frequency of back injuries when moving the standard seventy-pound mail sacks." If these blacks and women have accurate knowledge of the greater risks they face in particular circumstances, then they will use the market mechanism in an efficient way and will then select the job for which they are the most suited.[19]

As is probably evident, the assumptions underlying the market-efficiency argument are quite similar to those supporting the welfare argument. Both approaches require one to assume that employees' *preferences* will operate so as to attain authentic worker *welfare*. Both contain the implicit assumption that market-based preferences are accurate indicators of legitimate values. As has already been seen, however, this assumption is not generally true. If it were, there would never be grounds for governnment intervention in markets, e.g., to set minimum standards for work-place conditions. Likewise, were this assumption true, then one would have to condone the sweatshop conditions of a century ago. One would have to agree that twelve-hour workdays of a bygone era were efficient, because they allowed workers to choose an "efficient matchup". On the contrary, the efficiency and the optimality of worker choices, whether among anemia-prone Blacks or backache-prone women, is in part a function of the choices *available* to workers. If an economy is not diversified, and if employees have no real occupational alternatives in the face of the need to feed their families, then it can hardly be said that the "market . . . will promote efficient matchups".

The market-efficiency argument is also highly questionable in that the ethical conditions necessary for desirable market transactions are frequently not met. Recall that Viscusi maintained that (italics mine): "*If the worker bears all of the hazard associated with the risk and if he is cognizant of his own particular risk*, not simply the average risk for all, he will select his job optimally" with respect to his own risk potential and personal advantages and disadvantages. This means that, on its proponents' own terms, the validity of the market-efficiency argument is premised on workers' having adequate knowledge of their particular risk situations. But are people generally aware of their own risk potential? Most risk assessors would probably say that they are not. Starr, Whipple, and other proponents of the method of revealed preferences, as well as Fischhoff, Slovic, Lichtenstein, and other advocates of the method of expressed preferences (see note 10) have pointed out, repeatedly, that intuitive or subjective assessments of risks made by educated laymen are quite divergent from analytical, allegedly objective assessments of risks made by risk experts. Laymen typically overestimate low-probability risks and underestimate higher probability ones. For example, they overestimate catastrophic nuclear accident risks, but underestimate risks associated with automobile accidents.[20] Economists also realize that the public's risk perceptions are rarely accurate. To bridge the gap between the theoretical model of rational choice and actual, imperfect, real-life choice, economists almost always write the costs of searching for risk information into their equations dealing with choice under uncertainty. If these economists and risk assessors are correct, then the conditions necessary for ethical use of the argument from market efficiency are frequently not met. But if these conditions are not satisfied, then the argument does not provide convincing grounds for supporting the theory of the compensating wage differential.

2.2.3. *The Autonomy Argument*

A third reason for risk assessors' supporting the theory of the compensating wage differential is their allegation that it provides for more worker freedom and autonomy than would a theory not based on a monetary differential, but based instead on uniform standards. As one proponent of the autonomy argument puts it: "uniform standards do not enlarge workers' choices; they deprive workers of the opportunity to select the job most appropriate to their own risk preferences" and they enable rich persons to impose their risk preferences on lower income classes.[21] On this theory, acceptance of uniform risk standards and rejection of the compensating wage differential are not desirable because they represent "interference with individual choices".[22]

Like the previous two arguments, this one is also based on the doubtful presupposition that freedom and autonomy are served by identifying occupational *preferences* with authentic worker *welfare*. As has already been noted, such an identification does not work in all cases. The presupposition also fails to take account of the fact that, just because one holds a particular job, this does not mean that his occupation is an expression of his preferences. Many people engage in a certain work, not because they freely and autonomously choose to do so, but because they have no other alternatives. Moreover, in the absence of minimum standards for occupational safety, and in the absence of alternative opportunities for employment, one could hardly claim that his occupation was a result of autonomous choice. In fact, minimum risk standards or stricter safety requirements might actually enhance occupational autonomy, in the sense that workers might not be forced by circumstances to accept jobs whose risks were higher than they wished to bear. In failing to take account of the numerous factors which limit free choice, Viscusi and other proponents of the autonomy argument appear to assume, erroneously, that government safety regulations always limit workers' freedom, and that these alleged limitations are more significant than those imposed by lenient standards governing occupational safety.

2.2.4. *The Exploitation Avoidance Argument*

Many proponents of the theory of the compensating wage differential realize, however, that occupational safety and worker welfare are not always guaranteed simply by letting market forces operate. They know that often employees can be exploited by employers who are not forced to provide a safe working environment. To counteract this tendency, some proponents of the theory of the compensating wage differential maintain that a necessary condition for ethical implementation of this theory is that workers have adequate information about the risks they incur. According to Viscusi, "the most salient" form of market failure is inadequate worker information. "If workers and firms are not fully cognizant of the job risks resulting from their decisions, the desirable properties usually imputed to market outcomes may not prevail."[23] To avoid worker exploitation and market failure of the theory of compensating wage differentials, proponents of the theory often advocate employee education. Their view is that, once worker education is adequate, then market forces will drive compensating wage differentials so that optimal matchups between employees and occupations will occur.

Admittedly this exploitation avoidance argument is an improvement over arguments which ignore the role of occupational risk education but which support the theory of the compensating wage differential. Its flaw is in its major presupposition that education and compensation, alone, provide sufficient grounds for worker consent and autonomy. It takes too simplistic a stance as to the requirements for legitimate consent and free choice. As the torture example cited earlier reveals, other factors, besides one's knowledge of a situation and his being compensated for losses, determine the moral quality of his and others' choices about that situation. As was already noted, even a perfectly informed worker, who consented to the level of compensation for his high-risk job, nonetheless might have been forced to take the work, particularly if there were no alternative employment opportunities available or if he needed the money. This suggests that, in addition to workers' having full knowledge of their risk situation and being compensated for it, occupational choices must also be made in a context of ethically desirable background conditions. Such background conditions might include the operation of a free market and the existence of alternative employment opportunities. Without these background conditions, it is not clear that ethically desirable employee–employment matchups will occur.

Take, for example, the ethical desirability of choices made by miners who choose to work in Appalachian coal fields. (Appalachia includes much of the states of Kentucky, West Virginia, Virginia, Tennessee, North Carolina, and South Carolina.) It is well known that mining is one of the highest-risk occupations, that poorer workers are typically employed in the most risky jobs,[24] and that residents of Appalachia generally have no alternative to working in the mines, unless they want to move out of the region. This is because the Appalachian economy is not diversified, because there is no job training in a variety of jobs, and because absentee corporations who control 80 percent of all Appalachian land and mineral rights also control the only jobs; the situation is one of monopsony, where owners of most of the land also control most of the jobs.[25]

Even if Appalachian coal miners were generously compensated, and even if they all had perfect information as to the dangers of their jobs, the background conditions in the Appalachian economy would prevent their making a wholly voluntary choice to work in the mines. But if they were not able to make wholly voluntary choices as to the form of their employment, then it is not clear that proponents of the theory of the compensating wage differential can argue either that, since workers were aware that their jobs were extremely risky, those risks were freely chosen, or that the prevailing double standard with respect to occupational and public risks is acceptable to workers. In fact, if background conditions necessary for procedurally just choices (about forms of employment) are not met, it is not clear that implementation of the theory of the compensating wage differential is just. As John Rawls put it, "only against the background of a just basic structure . . . and a just arrangement of economic and social institutions, can one say that the requisite just procedure [for occupational and other choices] exists."[26] Sound as this insight about background conditions is, many risk assessors often neglect it in their considerations. In an otherwise excellent book on risk, even the brilliant philosopher Nicholas Rescher appears to neglect the role of background conditions in determining ethically acceptable risk choices. He speaks, for example, of suicide as being a "wholly voluntary" mode of death and of incurable disease as being a "wholly involuntary" mode of death.[27] Such language, however, ignores the importance of background conditions in determining what is more or less voluntary. Death by suicide might not be "wholly voluntary" (as he says) if it is a consequence of depression-induced medication whose side effects were unknown by the patient and by the doctor prescribing it. Likewise, death by incurable disease might not be "wholly involuntary" (as he says) if it is brought on more quickly by a person's unwillingness to take proper medical treatments, follow prescribed diets, etc. In other words, the line between what is voluntary and involuntary is quite uncertain in numerous cases. To the degree that philosophers and risk assessors ignore the numerous ways in which background conditions can affect the voluntariness of an action, to that same extent are they also likely to misjudge the voluntariness with which persons (e.g., Appalachians) choose to accept a particular level of risk. And to the degree that they misjudge voluntariness, they are also likely to propose inadequate theories about the ethics of risk acceptability.

In addition to the Appalachian example, there is further evidence for the thesis that, even with full information about risk, workers often are unlikely to make wholly voluntary decisions to accept high-risk employment situations. This is that people who can afford to do so generally avoid working in hazardous occupations. It is well known that, as a person's income increases, his willingness to accept risky situations decreases.[28] If this wealth–risk relationship holds, then workers' acceptance of high occupational risks is explicable by the constraints imposed by their low income and limited job skills, regardless of whether they understand the dangers to which they are exposed or not.

Even if proponents of the exploitation avoidance argument are correct in believing that proper education of workers can theoretically block exploitation of employees in high-risk occupations, it is still not clear that, practically speaking, such education can be accomplished to the degree necessary in all situations. In other words, even if education were a sufficient condition for insuring that high-risk workers voluntarily accepted the terms of their employment, it is not clear that this condition could be met in most situations. Hence, it is not clear that one would be justified in implementing a system of compensating wage differentials.

Why might the condition on education not be met? One reason is that either deliberately or out of negligence, companies and regulators have often kept their research findings about hazards secret from employees exposed to them. In the case of vinyl chloride, for example, long before workers were discovered to be at risk from liver cancer, there was strong enough evidence to support a presumption of a serious occuptional hazard. Similarly, decades after countries such as Japan have banned certain carcinogenic dye ingredients from the workplace, American workers "are still literally sloshing in them".[29] When company doctors have been aware of employment-induced illness, e.g., from asbestos in the Johns-Manville factory in Pittsburg, often they have covered up this fact for decades.[30]

Even some proponents of the compensating wage differential point out that "available evidence suggests that few firms make a comprehensive effort to inform workers of the risks they face. For example, no firms tell their employees the average annual death risk they face. Much information that firms do provide is not intended to enable workers to assess the risk more accurately .... Rather, it is directed at lowering workers' assessments of the risk. The most widespread claim by firms is that National Safety Council statistics indicate that the worker is safer at work than at home – a statement that ... is intentionally misleading [because some jobs are riskier than the average home, while others are not]."[31]

In situations where there is no deceit on the part of employers regarding the relevant risks faced by their employees, and in which workers are provided with full information, even this is not enough to insure that the practical conditions necessary for wholly rational occupational choices have been met. Proponents of the method of expressed preferences have pointed out that, even in the presence of complete disclosure on the part of the company, employees exposed to high-risk situations typically take on the "it won't happen to me syndrome".[32] The pervasiveness of this syndrome indicates that, even when the theoretical conditions for full employee education are met, they might not be satisfied in a particular concrete case, owing to misperception on the part of the worker. This in turn means that, because his knowledge is not operative, many employees likely are not making wholly voluntary decisions to work in high-risk situations.[33]

2.3. *Further Arguments*

In addition to these considerations that full education and compensation do not constitute sufficient conditions for affirming that employees in high-risk occupations accept their jobs in a fully voluntary sense, there are several other reasons why one might argue that the currently accepted theory of the compensating wage differential is not necessarily ethically defensible. One reason why even a very large wage differential might not offset the equity problems raised by the double standard for occupational and public risk is that the worker might not have the *right* to accept a particular risk.

2.3.1. *Acceptance of Some Risks Involuntarily Imposes Them on Others*

Consider the case in which a worker allegedly accepts a high workplace exposure to some carcinogen in exchange for a very high wage differential. The employee might be fully cognizant of the health hazards involved, and he might agree that the compensation afforded him is adequate. Nevertheless, he might not have the right to take the risk, perhaps because his acceptance of it inevitably puts other persons, who have not agreed to accept the risk, in jeopardy. Since most carcinogens are also mutagens, the worker who exposes himself to carcinogenic materials is also likely exposing his potential children and their descendants to mutagenic hazards. Of course, one might argue that unborn members of future generations have no rights to be protected from mutagenic risks taken by their ancestors, since they are not existent, and only existent beings have rights.

While the whole issue of rights of future generations is too extensive a topic to be discussed here,[34] one fact about the carcinogenic/mutagenic risk situation does seem clear. Workers might have a right to take a risk which threatens only themselves. It is less obvious that they have a right to take a risk which might damage something, the gene pool, which is beyond themselves. Hence it could well be questionable to assert that any person ever intending to reproduce has the right to accept workplace risks which are mutagenic when those risks are higher than those to which the public is normally exposed. As Rescher puts it so well: we can only take risks for ourselves, not for others; "morality enjoins conservatism".[35] Or, more strongly, the moral aspect of risk taking arises when the choices of individuals bear upon the interests of others.[36]

One does not have to move to future generations, of course, to discover innocent victims of some worker's alleged right to expose himself to industrial toxins in exchange for a higher wage. For example, some occupations, e.g., that of air-traffic controller, produce high psychological risks, e.g., that of stress. It might be questionable whether someone has the right to accept such a high-stress risk when the effects of the stress are not borne merely by the employee but also by his family. Likewise it might be questionable whether a particular worker, for example, in an asbestos factory, has the right to accept a higher workplace risk if such a risk might also affect his family. It is commonplace for the families of particular workers to contract cancer because they have been exposed to asbestos fibers carried home on clothing. Some wives have died of asbestos-induced cancer merely because they washed their husbands' clothing. Close contact with their fathers has also caused the children of asbestos workers to contract cancer, and recent U.S. examinations have revealed dangerous levels of lead in the blood of lead workers' children, chiefly as a consequence of inhaling lead dusts brought home on clothes.[37]

Admittedly some of the hazards faced by families of those in high-risk occupations could be eliminated or reduced by simple practices such as workers' bathing and discarding their work clothes before coming home. Nevertheless, to the extent that any employee's acceptance of a risk thereby places a higher health risk on someone other than himself, then to that same degree is his right to take such a risk highly questionable.

2.3.2. *Acceptance of Some Risks Is Based on Inconsistent Attitudes about Risk Perception*

A second reason why one might argue that a compensating wage differential does not justify societal acceptance of workplace risks which are higher than public risks is that proponents of the differential often defend their stance by making inconsistent appeals to the status of risk *perceptions* of workers. When Starr and other proponents of the theory of the compensating wage differential wish to justify workers' acceptance of higher risks in return for higher wages, they take an interesting stance. They maintain that, once employees are adequately educated regarding the risks they face, their risk *preferences* ought to be followed, and that regulators have no right to tell workers that they cannot follow their preferences for higher risks.[38] However, when these same proponents of the theory of the compensating wage differential wish to justify government imposition of particular standards for public risk, in the face of citizens' demands for stricter regulations, they take a different stance. They maintain that the risk preferences, even of highly educated laymen are subjective, intuitive, and generally erroneous, and therefore that regulators ought not merely to follow the public's demands for lower risks. Instead, they claim, regulators also ought to adhere to the risk assessments calculated by experts, since these reflect 'rational' preferences, and they ought to implement them in standards for public exposure.[39]

For example, speaking of the public's 'irrational' aversion to low-probability, high-consequence nuclear accidents, Starr and Whipple maintain that laymen's perceptions regarding this technology are incorrect and that their demands for greater nuclear safety are not reasonable, since they fly in the face of experts' beliefs about acceptable levels of nuclear risk.[40] Psychometric surveys of attitudes about risk reveal that there is no significant difference, in level of relevant technical knowledge, between those who favor greater nuclear safety for the public and those who favor less stringent nuclear risk standards.[41] Nevertheless Starr and others claim that the preferences of the public for lower nuclear risks ought not to dictate risk standards.[42]

Proponents of the theory of the compensating wage differential, who *advocate* adherence to worker perceptions of risk, in order to justify less stringent *occupational standards*, are thus in an apparently contradictory position when they *condemn* adherence to risk perceptions of relevantly educated laymen, in order to justify their proposals for less stringent *public standards*. They cannot have it both ways. Either government standards ought to be based on the risk perceptions of adequately informed persons who are likely to be affected by the risk, or they ought not to be based on these risk perceptions. If risk assessors claim that relevantly educated persons err in their risk perceptions and ought to be 'corrected' by experts, then both workers and the public ought to be so corrected, and not just the public.

Of course, the main objection to this appeal for consistency, in valuing the risk perceptions of those who are adequately informed about a particular hazard, is that the cases of worker perceptions and public perceptions are not analogous. One might argue, following this line of objection, that workers voluntarily accept given modes of employment and that they have differential compensation for the risks they face, whereas the public has neither. Because of the alleged consent and compensation involved in the worker case, so the objection goes, workers' preferences ought to be followed, whereas risk preferences of the public (which generally involve cases of less consent and no compensation) need not be followed.

As this objection correctly notes, the cases of worker's perceptions and the public's perceptions are disanalogous with respect to consent and compensation. It does not follow, however, that these disanalogies are morally relevant in justifying inconsistent treatment of risk perceptions. Why not? If the fact that one is compensated for the risk he chooses is reason to follow his risk preferences, as proponents of the compensating wage differential argue, then the fact that one is *not* compensated for the risk imposed on him is even greater reason to follow his preferences for lower risks. Virtually all risk assessors, especially those who adhere to the method of revealed preferences, including Starr and other advocates of the compensating wage differential, maintain that risks taken voluntarily are more acceptable than risks of the same level which are involuntarily imposed.[43] But if this is so, then there is greater reason to follow public preferences for lowering risks to which citizens are involuntarily exposed than there is for following worker preferences regarding risks for which consent is not necessarily wholly voluntary (see Section 2.2.4 in this chapter). In other words, the very disanalogies between worker risk and public risk, with respect to compensation and consent, indicate that, if anything, there is very likely more reason to follow public preferences for lower risks than to follow worker preferences for higher risks (since the public is neither compensated for societal risks nor given a choice as to whether to accept them, and since workers' acceptance of jobs is often not voluntary, owing to questionable background conditions). Yet, this is exactly the *opposite* of the view taken by most risk assessors, who argue against following societal preferences for lower risks and in favor of worker preferences for higher risks. This means that proponents of the compensating wage differential are on shaky ground when they both reject and accept risk preferences, depending on whether those preferences emerge from the public or from workers. If Starr, Whipple, and other risk assessors are correct in rejecting public preferences about societal risks, then it is highly questionable to invoke worker preferences in order to help justify the theory of the compensating wage differential.

3. CONCLUSIONS AND ALTERNATIVES

If these responses to the arguments in favor of the theory of the compensating wage differential are correct, then it is likely that appeal to this theory is not adequate grounds for defending a double standard with respect to occupational and public risks.

Compensation and allegedly voluntary choice of occupation do not guarantee that a particular level of worker risk is ethically acceptable, any more than compensation and the subject's consent guarantee that other actions affecting a subject are ethically acceptable. As was already pointed out, if a particular action is wrong, e.g., engaging in nontherapeutic experimentation on human beings, then the fact that the human beings may have consented to the experimentation and that they may be compensated for it, does not change the ethical quality of the act (of experimentation) from 'undesirable' to 'desirable'. Admittedly, however, consent and compensation might render the act less undesirable than it might have been, were the subject not to have given consent or not to have received compensation. In other words, the whole question of whether a double standard for occupational and public risk is morally acceptable cannot be reduced simply to the issues of compensation and consent, as proponents of the theory of the compensating wage differential appear to do.

If compensation and consent are not the only relevant considerations in deciding whether the double standard for occupational and public risk is acceptable, then the theory of the compensating wage differential, alone, does not provide grounds for accepting such a double standard. This means that, in the absence of some ethical justification for the double standard, the best policy might be to follow a principle of *prima facie* equality.[44] In other words, in the absence of good reasons for discriminating with respect to equal protection, the best policy might be to protect persons equally, whenever possible, regardless of whether they are workers or members of the public.

If it turns out that there are compelling reasons, other than the theory of the compensating wage differential, for continuing to maintain a double standard with respect to occupational and public risk, then those reasons will need to be defended. One place to begin investigating whether there are good reasons for maintaining a double standard might be to think of worker risk as analogous to patient risk. Although there is an ethical and a legal requirement for informed consent on the part of a patient being treated by a medical doctor, one of the limitations of the current policy of following the theory of the compensating wage differential is that there is no legal requirement for informed consent in the workplace. One might well argue, however, that just as persons now claim that a doctor's withholding information from a patient is a violation of the medical doctor's fiduciary role and a way of undermining the patient's autonomy, so also it might be said that an employer's withholding risk information from employees is a violation of the employer's fiduciary role and a way of undermining the employees' autonomy. Were there a recognized ethical and legal requirement for informed consent in the workplace, then the case for the ethical desirability of the compensating wage differential would be much stronger. Indeed, with an adequate means of insuring informed consent, and with a situation of just background conditions, there might be ethical grounds for justifying a double standard with respect to occupational and public risk.

Regardless of possible future justifications of the current double standard for risk, one thing is certain. The theory of the compensating wage differential, as now implemented, does not adequately safeguard worker autonomy and well-being, for all the reasons spelled out earlier. This means, at a minimum, that risk assessors ought either to accept uniform risk standards, on grounds of *prima facie* equality, or to provide further ethical justification for their current adherence to double standards. If risk assessors do neither, then they will be open to the charge that they have allowed marked expediencies, rather than ethical convictions, to dictate standards for worker health and safety.

NOTES

1 Carl Gersuny, *Work Hazards and Industrial Conflict*, University Press of New England, Hanover, New Hampshire, 1981, p. xi; hereafter cited as: *WHIC*.

2 Gersuny, *WHIC*, p. 1.

3 M. Douglas and A. Wildavsky, *Risk and Culture*, University of California Press, Berkeley, 1982, p. 9; hereafter cited as: *RAC*.

4 W. K. Viscusi, *Risk by Choice*, Harvard University Press, Cambridge, 1983, pp. 114–115, 136; hereafter cited as: *RBC*.

5 E. Eckholm, 'Unhealthy Jobs', *Environment,* **19** (6), (August/September 1977), 31–32; hereafter cited as: Jobs.

6 Quoted by Eckholm, Jobs, p. 32. For an excellent treatment of the history of occupational risk and disease, see D. M. Berman, *Death on the Job*, Monthly Review Press, London, 1978; hereafter cited as: *DOJ*. See also the numerous case studies in *Quantitative Risk Assessment in Regulation* (edited by L. B. Lave), Brookings Institution, Washington, D.C., 1982, Chapters 3–8; hereafter cited as: *QRA*.

7 R. W. Kates, *Risk Assessment of Environmental Hazards*, John Wiley and Sons, New York, pp. 46–47; hereafter cited as: *RA*.

8 Berman, *DOJ*, pp. 192–193; Kates, *RA*, pp. 168–174.

9 *Code of Federal Regulations*, 10, Part 20, U.S. Government Printing Office, Washington, D.C., 1983, p. 235; see also pp. 230–241.

10 See Sections 3.3.2 and 3.3.3 in Chapter Two of this volume.

11 For this point of view, see C. Starr, 'Social Benefit Versus Technological Risk', *Science* **165** (3899), (19 September 1969), 1232–1233; hereafter cited as: Social Benefit. See also N. Rescher, *Risk: A Philosophical Introduction*, University Press of America, Washington, D.C., 1983, p. 172, who argues that involuntary risks are less acceptable and hence ought to be subject to more stringent standards; hereafter cited as: *Risk*.

12 Starr, 'General Philosophy of Risk–Benefit Analysis', in *Energy and the Environment* (edited by H. Ashley, R. Rudman, and C. Whipple), Pergamon, New York, 1976, p. 16; hereafter cited as: Philosophy and *EAE*. See note 10 above.

13 Viscusi, *RBC*, p. 38.

14 Viscusi, *RBC*, pp. 37 ff., 156–168. See also Douglas MacLean, 'Risk and Consent: A Survey of Issues for Centralized Decision Making', working paper, Center for Philosophy and Public Policy, University of Maryland, College Park, Maryland, 1981, pp. 6–9; MacLean refers to the theory of compensating wage differentials as part of what he calls the "model of implied consent".

15 Viscusi, *RBC*, pp. 46, 52.

16 Viscusi, *RBC*, p. 52. See also Wildavsky and Douglas, *RAC*, pp. 69 ff.

[17] See the discussion of preferences versus values in Section 3.3.3 of Chapter Two.
[18] Viscusi, *RBC*, pp. 132, 135.
[19] Viscusi, *RBC*, pp. 132–133.
[20] See, for example, C. Starr and C. Whipple, 'Risks of Risk Decisions', *Science,* **208** (4448), (6 June 1980), 1115–1117; hereafter cited as: Risks. See also B. Fischhoff, P. Slovic, and S. Lichtenstein, in *Societal Risk Assessment* (edited by R. Schwing and W. Albers), Plenum, New York, pp. 192, 202, 208; hereafter cited as: *Risk*. Finally, see B. Fischhoff, P. Slovic, S. Lichtenstein, S. Read, and B. Combs, 'How Safe Is Safe Enough?' *Policy Sciences,* **9** (2), (1978), 140–142, 148–150; hereafter cited as: Fischhoff, Safe.
[21] Viscusi, *RBC*, p. 80.
[22] Viscusi, *RBC*, p. 83.
[23] Viscusi, *RBC*, p. 76; see also pp. 77–87.
[24] See, for example, M. W. Jones-Lee, *The Value of Life: An Economic Analysis*, University of Chicago Press, Chicago, 1976, p. 39; Eckholm, Jobs, pp. 33–34. See Starr, Philosophy, pp. 15 ff., and Viscusi *RBC*, p. 46.
[25] John Egerton, 'Appalachia's Absentee Landlords', *The Progressive,* **45** (6), (June 1981), 43–45, and J. Gaventa and W. Horton, Appalachian Land Ownership Task Force, *Land Ownership Patterns and Their Impacts on Appalachian Communities*, Vol. 1, Appalachian Regional Commission, Washington, D.C., 1981, p. 25–59, 210–211.
[26] J. Rawls, *A Theory of Justice*, Harvard University Press, Cambridge, 1971, p. 87.
[27] Rescher, Risk, p. 173.
[28] See, for example, B. A. Emmett, *et al.*, 'The Distribution of Environmental Quality', in *Environmental Assessment* (edited by D. Burkhardt and W. Ittelson), Plenum, New York, 1978, pp. 367–71, 374, and P. S. Albin, 'Economic Values and the Value of Human Life', in *Human Values and Economic Policy* (edited by S. Hook), New York University Press, New York, 1967, p. 97. See also M. Jones-Lee, *The Value of Life*, University of Chicago, Chicago, 1976, pp. 20–55.
[29] Eckholm, Jobs, p. 33.
[30] Berman, *DOJ*, pp. 1–4.
[31] Viscusi, *RBC*, p. 71.
[32] Starr, Philosophy, p. 5.
[33] See Viscusi, *RBC*, pp. 60–75.
[34] For discussion of this topic and relevant bibliographical materials, see K. S. Shrader-Frechette, *Environmental Ethics*, Boxwood, Pacific Grove, Ca., 1981, Ch. 3.
[35] *Rescher*, Risk, p. 161.

36 *Rescher*, Risk, p. 162.

37 Eckholm, Jobs, p. 30.

38 Viscusi, RBC, pp. 77, 80, 83. See Starr, Social Benefits, pp. 1233–1234, and Starr, Philosophy, pp. 15–21.

39 Starr and Whipple, Risks, pp. 1115–1119.

40 Starr and Whipple, Risks, pp. 1115–1117, esp. p. 1117.

41 Fischhoff, Safe, p. 150. See also note 20.

42 See Chapter Six (and especially Section 4.) of this volume.

43 Starr, Social Benefit, pp. 1233–1234; Starr, Philosophy, pp. 26–30.

44 See Chapter Three of this volume for a discussion of the principle of *prima facie* equality.

PART THREE

DECISION-THEORETIC PROBLEMS WITH THE METHOD OF REVEALED PREFERENCES

CHAPTER FIVE

RISK EVALUATION AND THE PROBABILITY-THRESHOLD POSITION

1. INTRODUCTION

In a recent article in *American Scientist*, a scientist and public policy expert quipped: "Chicken Little is alive and well in America".[1] Never in history have health and environment-related hazards been so low, he said, while "so much effort is put into removing the last few percent of pollution or the last little bit of risk".[2] His thesis, like that of many others, was that our criteria for defining 'acceptable risk' often are too stringent.

Are the criteria for acceptable risk too stringent? Those who believe that they are, especially advocates of the method of revealed preferences,[3] often maintain that society ought to ignore very small risks, i.e., those causing an average annual probability of fatality of less than 10^{-6}. I call this view (1) the 'probability-threshold' position. Opposed to this stance, other participants in the risk-evaluation debate hold views which may be described as (2) the zero-risk position and (3) the weighted-risk position.

The zero-risk position, (2), is the belief that a risk is acceptable only if it has no measurable negative effect on human health or well-being. This stance is held by numerous environmental activists and has been popularized recently by the well-known antinuclear scientists, John Gofman and Arthur Tamplin.[4] The best-known example of legislation which presupposes this position is the so-called 'Delaney Amendment' to the Pure Food and Drug Act, passed by Congress in 1958. It states that "no [food] additive shall be deemed 'safe' if it is found . . . after tests that are appropriate for the evaluation of the safety of food additives to induce cancer in man or animals".[5] (The purpose of the Delaney Amendment is to prohibit the *intentional* addition of chemicals to food if those chemicals cause cancer; "inadvertent contaminants", or those not intentionally added, are dealt with in Section 406 of the Federal Water Pollution Control Act. The Food and Drug Administration weighs various factors (health, economics, supply of the food, for example) in determining what levels of inadvertent contaminants are acceptable.[6])

Although the zero-risk position is not the focus of this discussion, very likely one could argue that it is not a desirable basis, in all situations, for determining levels of acceptable risk. For one thing, it ignores the importance of trade-off; perhaps certain nonzero risks ought to be judged acceptable because of the benefits their acceptance might bring. *Second*, the view presupposes that there are zero-cost alternatives in every risk decision, a thesis which might well be shown incompatible with the facts. *Third*, the position presupposes that there will not be scientific progress, such that smaller and smaller amounts of harmful substances always can be detected in virtually every substance. If this presupposition is incorrect, as advances in analytical chemistry suggest, then the zero-risk position would require prohibitions against innumerable substances – prohibitions which would likely be impractical.

The third view, which I have called the weighted-risk position, is built on the assumption that the most acceptable risk is that which is determined by an ethically weighted risk-cost-benefit analysis. As defended by Kneese, Ben-David, and Schulze, this position relies on weighting the various risks, costs, and benefits according to alternative ethical criteria.[7] Since this view has been analyzed elsewhere,[8] I will not discuss it here.

The first position, the probability-threshold view, starts from an assumption denied by those in the zero-risk camp, viz., that a certain amount of risk is acceptable. Instead, say proponents of the threshold thesis, one ought to ignore any small risks (those, for example, for which the individual probability of fatality is at the threshold of 10^{-6} per year or less) and, above this level, one ought to choose the risk (as most acceptable) which has the lowest probability of a unit cost burden (e.g., annual individual probability of fatality) occurring. A 1977 NAS (National Academy of Sciences) report urged that this position regarding risk be adopted whenever possible.[9]

Many regulatory agencies, including the Environmental Protection Agency (EPA) and the Nuclear Regulatory Commission (NRC), as well as numerous scientists, engineers, policymakers, and environmental assessors, such as Comar, Hushon, Okrent, and Gibson have argued forcefully for this probability-threshold view.[10]

Of the three risk theories just mentioned, the probability-threshold position is by far the most dominant in public policymaking. Because of its singular importance, and because I believe it is fundamentally erroneous, I will focus on the arguments which have been used to support this view. Although the analysis is not complete, I hope to investigate some of the major arguments used on behalf of the probability-threshold position. It should become clear that this view rests on some highly questionable mathematical, ethical, ontological, and epistemological presuppositions.

2. THE PROBABILITY-THRESHOLD POSITION

Proponents of the 'de minimus' or 'probability-threshold' view[11] claim that hazards whose probabilities (usually, average annual probabilities of fatalities) are below a given level, ordinarily 10^{-6} per person, are unimportant or insignificant from the point of view of the risk assessor.[12] Okrent, for example, argues that risks ought to recieve attention proportionate to their probability and magnitude.[13] He claims, as a consequence, that there is a "risk threshold below which no review" or risk assessment ought to be required.[14] Likewise, Gibson claims that assessors should use a "common approach to all hazards" by reducing them all to one "socially acceptable level" of risk.[15] Starr and Whipple repeatedly affirm that this threshold, the "lower limit for concern about risk", is "the natural-hazards mortality rate", which is approximately 10^{-6} per person per year.[16]

The authors of the most comprehensive study of commercial nuclear reactor safety, U.S. WASH-1400, known as the Rasmussen Report, used precisely this figure as a threshold below which risk was considered to be negligible. The authors of the NRC report affirm, for example, that "an accident fatality risk to the public of . . . 10^{-6} or lower is considered negligible."[17] They also claim that, in their report, "the highest level of acceptable risk has been taken as the normal U.S. death rate from disease."[18] Next they argue that nuclear risk probabilities fall within this range, since "the operation of 100 reactors will not contribute measurably to the overall risks due to acute fatalities and property damage from either man-made or natural causes."[19] The authors also support taking the nuclear risk, they say, because "we do not now, and never have, lived in a risk-free world" and because "nuclear accident probabilities are so low that they are insignificant compared to the overall risks due to other man-made and natural risks."[20]

Subsequent U.S. Nuclear Regulatory Commission (NRC) directives, promulgated in January 1983, have defined risks smaller than 10^{-6} as negligible and therefore acceptable. Moreover, as Rescher points out, "as a matter of practical policy we [decisionmakers setting risk standards] operate with probabilities on the principle that when $x < \epsilon$, then $x \cong 0$. Where human dealings in real-life situations are concerned, sufficiently remote possibilities can – for all sensible purpose – be viewed as being of probability zero, and possiblities with which they are associated set aside . . .

a certain level of sufficiently low probability can be taken as a cut-off point below which we are no longer dealing with '*real* possibilities' and with '*genuine* risks'."[21]

2.1. *Main Arguments in Favor of the Probability-Threshold View*

Justification of this threshold position takes several forms, which may be organized in terms of the types of reasoning to which risk assessors, philosophers, decision theorists, and policymakers appeal, I call these:

(1) the argument from decision theory
(2) the argument from ontology
(3) the argument from epistemology
(4) the argument from economics
(5) the argument from psychology
(6) the argument from history; and
(7) the argument from politics.

In this discussion, I will clarify and criticize the first three of these arguments, since these appear most closely related to philosophical and methodological considerations, and since these arguments appear to dominate the policy debate over the probability-threshold position. First, however, let me describe several of the types of arguments belonging in the four latter categories.

As made by risk assessors, scientists, engineers, and environmental policymakers, (4) the argument from economics is quite persuasive. In one variant of it, which I shall call (4A), assessors claim that a threshold (below which risks are considered negligible) is economically necessary and desirable, because funds spent to reduce risk cannot be used to increase industrial productivity.[22] A related argument, which I shall call (4B), is that a probability threshold is necessary and desirable because the costs of not recognizing such a cut-off point often exceed those of recognizing it. For example, says Wildavsky, eliminating the subthreshold risk from pesticides by banning them may result in less food and higher food prices in countries that import from us. Likewise, eliminating the subthreshold risk from nuclear fission by banning atomic generation of electricity may result in flooding Indian lands to create dams in Canada.[23]

Proponents of (5) the argument from psychology often use one of at least two claims to justify their adherence to the threshold position. They claim, in what I call (5A), that the public is not opposed *a priori* to bearing *subthreshold* risks but is opposed to bearing *new* ones and that, as a consequence, one cannot argue that it is rational or desirable, *a priori*, to eliminate all subthreshold risks.[24] A related claim, which I shall call (5B), is that those who refuse to accept subthreshold risks have "pathologic fear". or "near clinical paranoia", or a subnormal level of trust and confidence.[25]

Another popular line of justification is (6) the argument from history. Proponents, of what I call claim (6A), maintain that society has followed a threshold notion in the past and it has led to prosperity; without society's having done so, they claim, there would be no airplanes, no mechanization, and no high technology. Therefore, they say, this past practice of accepting the probability-threshold position should be continued.[26] A related claim, which I call (6B), is that since society has not been hurt in the past by following the probability-threshold view, it will not be hurt by doing so in the future, because people have been, and always will be, able to accommodate themselves to new developments.[27]

Variants of (7), the argument from politics, take at least three forms. One is (7A), that since more risks come from individually chosen behavior (like smoking or drinking), rather than from industrially or societally imposed hazards (like nuclear power), and since nearly all societal risks are in the subthreshold category; lawmakers should strive to eliminate individual risks before industrial-societal risks; hence, adherence to the threshold position is desirable and justifiable until all higher-level or personally chosen risks are reduced to the level of most subthreshold risks.[28]

A quite different variant is (7B), the argument that one ought not to ask the government to regulate risks below a given threshold because such regulation merely increases undesirable governmental restrictions.[29] According to yet another version, (7C), acceptance of the threshold position is necessary in order to avoid "the nuisance aspects of regulation".[30]

Because they are more specifically 'philosophical' (rather than social-scientific) than arguments (4)–(7), and because analysis of them requires less lengthy inquiry into largely *factual* (as opposed to normative or theoretical) concerns than does inquiry into (4)–(7), I will focus my discussion on arguments (1)–(3). These are, respectively, the arguments from decision theory, ontology, and epistemology.

2.2. *The Argument from Decision Theory*

The argument from decision theory is widely held by a number of decision theorists, act utilitarians, and risk assessors (especially those who subscribe to the method of revealed preferences). Act utilitarians and decision theorists both attempt to maximize *expected utility*. (This is the sum of possible benefits, minus the possible risks and costs, where each benefit, risk, or cost is multiplied by the probability that it will occur.) As a result of calculating an event's expected utility, proponents of the argument from decision theory maintain that very low probabilities (those smaller than 10^{-6} annual probability of fatality, for example) ought to be ignored because they are negligible.

Allen Buchanan, Frank Miller, and Rolf Sartorius, for example, all argue that the act utilitarian ought to ignore very small probabilities. Discussion the problem of maximizing group utility,

Buchanan maintains that a rational person ought to pay no attention to very small chances.[31] Taking a similar position, Miller and Sartorius sanction what they call "the prevailing view".[32] As stated by Mancur Olson,[33] this is the belief that any rational person would ignore a probability which was not "perceptible"; for example, any rational person would not contribute to the provision of a public good if his individual contribution were imperceptible.

Although they do not explicitly invoke the concept of expected utility, a number of other theorists also appear to employ a variant of the argument from decision theory; they maintain that rational persons ought to ignore small probabilities because they are "remote ... not *real* possibilities", negligible, or not perceptible. Nicholas Rescher, for example, maintains that "sufficiently remote possibilities can – for all sensible purpose – be viewed as being of probability zero, and possibilities with which they are associated set aside."[34]

2.2.1. *Origins of the Argument from Decision Theory*

The tendency of philosophers, risk assessors, and decision theorists to ignore low-probability events below a certain threshold has a number of historical precedents. Cournot, d'Alembert, Condorcet, and others used variants of what I call 'the argument from decision theory' when they attempted to justify neglect of extremely small probabilities. One of the most famous of these justifications was given by Buffon in connection with the St. Petersburg Paradox.

The 'St. Petersburg Paradox' arises in an imaginary gamble, the St. Petersburg game, which has played a key role in the history of uncertainty theory. To understand the paradox, recall that the earliest mathematical studies of probability were concerned with gambling, especially with the question of which of several cash gambles would be most advantageous. Early probability theorists maintained, according to the 'principle of mathematical expectation', that the gamble with the highest winnings would be best, or in terms of utility, that wealth measured in cash is a utility function.[35] One of the many ways in which this principle was justified was by arguing that it follows for the long run from the weak 'law of large numbers', as applied to large numbers of independent bets, in each of which only sums, that the gambler considers small, are capable of being won or lost. In its simplest form, due to Jacob Bernoulli, the law states that, in a sequence of *independent* trials, in each of which a given event E may occur with a constant probability p, the probability, that the relative frequency of occurrence of E in n trials differs from p by more than any assigned positive quantity, can be made as small as desired by making n sufficiently large. Even though, in any *finite* number of trials, no matter how large, relative frequency cannot be identified with true probability, Buffon and Cournot have suggested, "as a general principle . . . events whose probability is sufficiently small are to be regarded as morally impossible."[36] They used this principle to resolve the 'St. Petersburg Paradox'.

To state the paradox succinctly, suppose that one must determine how much he should pay to play a game in which infinite winnings are expected. The game is played with a fair coin which is tossed n times, until a head appears. The individual is then paid 2^n dollars. Since the mathematical expression $\Sigma_{n=1}^{\infty} (2^n)(1/2n)$ is infinite, the mathematical expectation of winnings is infinite and, says Arrow, one " 'should' be willing to pay any finite stake", any number of dollars, in order to play the game.[37] The paradox arises because intuition tells us that there are some finite amounts which are so large that most people would not pay such prices, and "this 'fact' will have to be explained".[38] Buffon explains this fact by appealing to the principle of neglecting extremely small probabilities. "The probability that a head will not appear until the nth toss becomes very small for n sufficiently large; if the occurrence of that event is regarded as impossible for all n beyond a certain value" or probability threshold, then the mathematical expectation of return becomes *finite*, and the paradox is resolved.[39] In other words, Buffon's argument in favor of assuming that small-probability events are impossible is that, if we make this assumption, then we can explain why reasonable persons would not pay an amount greater than a certain threshold, even to obtain infinite winnings. The analogous, contemporary policy argument used by philosophers and risk assessors is that, if we make this assumption, then we can explain why reasonable persons would not pay an amount greater than a certain threshold, even to obtain perfect health, safety, or environmental control (assuming that they could be had).

2.2.2. *Problems with the 'Argument from Decision Theory'*

There are at least two main problems with this 'argument from decision theory', whether articulated by persons like Buffon, or by contemporary risk assessors. *First*, the critical assumption, that certain low probabilities can be counted as zero, need not be made in order to explain public reluctance to pay either for infinite expected winnings or for perfect environmental conditions. *Second*, there is the practical problem of choosing which low probability is the critical threshold. Let us examine, in order, each of these difficulties.

First, the assumption that, below a given threshold, probabilities count as zero, need not be made in order to explain public reluctance to pay a large amount for infinite expected winnings. This is because the form of this inference ("if one assumes that, below . . . , then one can explain public . . .) is retroductive. As a consequence, a great many alternative assumptions could just as easily explain the public reluctance exhibited in the St. Petersburg Problem. What might some of these assumptions be?

First, one might avoid the reasonable or predictable tendency to pay a large finite amount for infinite expected winnings by assuming that, since n approaches infinity, and since he will not be alive for infinity, it makes no sense for him to spend an inordinately large amount in order to play a game whose winnings may not be realized in his lifetime. In other words, the 'paradox' arises only if the expected winnings are infinite. But the expected winnings are infinite only if n approaches infinity. The player need not assume that n approaches infinity and, in fact, so far as his lifetime is concerned, $n \neq \infty$. Hence, for the lifetime of the player, the mathematical expectation of return would be *finite*. But if it is finite, then the paradox would be resolved in a way not dependent upon the assumption that certain low-probability events are counted as zero. One might reason that, during his lifetime (or even during that of several generations of his descendants), the mathematical expectation of return would be *infinite*, and surely less than some other finite amount which might be required to play the game; hence it would not be reasonable to pay any finite amount in order to play the game.

Alternatively, apart from considerations are to whether $n \to \infty$, and as to whether mathematical expectation is infinite during the player's lifetime, one could make yet another assumption. This also is independent of the one that low probabilities count as 0, and it would also resolve the paradox. One might simply assume that some large finite amount, e.g., billions of dollars, could not be paid for each time the game was played, since the rationality of such payment would be a function of the available monies of the player.

This second alternative, to the assumption that certain low probabilities are zero, is very similar to (Daniel) Bernoulli's claim that persons of prudence do not invariably obey the principle of mathematical expectation. Interestingly, Bernoulli claimed neither that people assume that, below a certain threshold, very low probabilities are zero, nor that acceptance of this assumption is why people fail to obey the principle of mathematical expectation. Instead, he maintained that the dollar that might be precious to a pauper would be almost worthless to a millionaire.[40] In the language of economics, this assumption might be called 'the law of diminishing marginal utility'. Apart from whether there are cases in which it might not make sense to accept this assumption, it would nevertheless be reasonable to use it to explain much behavior regarding the St. Petersburg Paradox. A poor man certainly could pay less than a rich man, in order to play a game whose mathematical expectation was infinite. But if so, then there is no need to claim that it is necessary to make the assumption, that certain low probabilities are 0, in order to explain the paradox. One might just as well assume that marginal utility diminishes or that the rate of exchange between circumstances producing discomfort or well-being, and money, depends on the wealth of the person involved.[41]

A second problem with 'the argument from decision theory' is that of choosing which low probability is the critical threshold. Any choice in this regard seems to be purely *arbitrary* in specifying a particular level.[42] Buffon, for example, did not use the 10^{-6} threshold; he argued that a probability less than 10^{-4} is effectively zero. His reasoning was that no man of 56 in good health expects to die in one day, and that in 18th-century Europe that actuarial probability was approximately 1 in 10,000.[43]

One central problem with Buffon's reasoning, however, is why one's *belief* about his chances of dying constitute sufficient grounds for his *knowledge* that 10^{-4} is effectively zero. Moreover, as a mathematical claim, the thesis (that probabilities less than 10^{-4} are 0) is unarguably and obviously false. Arrow, in his *Essays in the Theory of Risk Bearing*, put the matter well. Given "an exhaustive set of mutually exclusive individual events, each of which has a probability less than critical value, it would be contradictory to say that all of them were impossible, since one must occur."[44] Brian Barry made a similar point; "if each contribution is literally 'imperceptible', how can all the contributions together add up to anything?[45]

As a claim about *social policy*, the thesis, that probabilities below 10^{-4} or 10^{-6} ought to be counted as zero, is question-begging. It is question-begging because it presupposes that the magnitude of some number, rather than factors such as the equity of distribution, risk–benefit trade-off, or compensation, determine the acceptability of a given risk. For example, the thesis, that probabilities below 10^{-4} or 10^{-6} ought to be counted as zero, appears plausible only if it involves no violations of justice, and if there are no other alternatives to accepting it. Gibbon, the historian of the Roman Empire, was said to have made the same point. Allegedly he replied to Buffon's proposal:

> If a public lottery were drawn for the choice of an immediate victim, and if our name were inscribed on one of the ten thousand tickets, should we be perfectly easy?[46]

If our obvious answer is 'no', then it appears that Buffon's thesis begs questions about justice and alternatives (to assuming that probabilities less than $10^{-4} = 0$). Because epsilon is not zero, any social-policy claim, that probabilities less than 10^{-4} ought to be treated as 0, always requires that one give reasons for such a proposal. In the absence of such reasons, one can always ask: "Why should I accept even a small increment of risk?"

2.2.3. *A Rejoinder: The Statistician's Case for the Argument*

To all these charges, risk assessors and decision theorists might make both practical and theoretical responses. Their *practical* response is likely to be that remote possibilities must be dismissed, or one will be faced with endless alternatives and stultifying actions.[47] This response will be discussed later. Their *theoretical* reply is likely to be that many eminent statisticians and economists nevertheless subscribe to 'the argument from decision theory'. Since it is reasonable to evaluate the case for this argument only after investigating the very best defenses of it, let us examine the case that an eminent mathematician and economist, Kenneth Arrow, makes on its behalf.

Arrow basically adopts the same line of defense for his view as that taken by K. Menger, the little-known German economist. Menger claimed that the discrepancy between actual risk behavior and that based on expected utility, or the mathematical expectation of return, is explained, in part, through undervaluation of small probabilities. (This was a generalization of the view that one ought to neglect small probabilities, as in the probability-threshold position.[48])

Arrow defends a generalization of Menger's viewpoint and argues that "there is some ordering of probability distributions" since one can assign "to each [member of a] probability distribution a number representing its desirability."[49] He calls such an assignment of numbers a "risk preference functional", and claims that "it is analogous to the assignment of numbers to commodity bundles in order of their desirability, which we call a utility function."[50] Hence, on this view, one's "risk preference functional" could be so arranged that any number of different probabilities, all below a certain threshold, could be said to be of equal desirability. One might define the desirability of a probability distribution in terms of various parameters, e.g., the size of the mean, the smallness of the standard deviation, or, in terms of insurance, the probability that income will fall below some critical level. Other authors have suggested that the desirability of a given distribution is a function of its *risk* and *yield*.[51]

Following the Von Neumann and Morgenstern principle, that

probability distributions are ordered,[52] Arrow argues that there is a method of *assigning utilities* [italics mine] to the individual outcomes so that the utility assigned to any probability distribution is the expected value under that distribution of the utility of the outcome. The numbers assigned are said to be utilities because they are in the same order as the desirability of the distribution or outcome (desirability itself being merely an ordinal concept)."[53] In other words, despite all his complex references (into which I have not gone here) to the work of Menger, Marschak, Hicks, Cramer, Domar, Musgrave, Keynes, Von Neumann, and Morgenstern,[54] Arrow has shown merely that, because there is a method of assigning utilities to individual outcomes, there is a way to order the elements of various probability distributions. He has argued neither for a *particular* ordering, such that all subthreshold probabilities are to count as zero, nor for the thesis that only a particular ordering can explain rational risk behavior.

Hence, Arrow has provided merely a *sufficient*, but not a *necessary*, condition for explaining human choices regarding safety in terms of neglecting small subthreshold risks. Moreover, in establishing that there is a way to assign utilities to individual outcomes so as to order the elements of various probability distributions, Arrow indeed has 'explained' the value attributed by some persons to low-probability risks. His explanation, however, has been bought at the price of separating the notion of 'utility' from that of 'good'. As he himself admits, "the utilities assigned are not in any sense to be interpreted as some intrinsic amount of good in the outcome . . . all the intuitive feelings [regarding utility] . . . are irrelevant."[55] This means that Arrow, who began discussing 'ordering' the elements of probability distributions, as a way to relate risk theory with human intuitions about risks and welfare, has ended by showing us that such orderings are possible, but only if they are *not related* to human intuitions about preference orderings being related to what is somehow good or desirable.

If Arrow's arguments are representative, and I think that they are, then they suggest that risk assessors cannot have it both ways. They cannot claim, *both* that some ordering theory justifies their counting subthreshold probabilities as zero, *and* that this ordering is necessarily related to what is good, preferable, or desirable in any real sense. In other words, Arrow's analysis, just as that of the risk assessors', presumes some criterion for ordering various probabilities. Since no nonarbitrary criterion is advanced by Arrow, his precise analysis does not help the case of proponents of the probability-threshold view. The same key problem faces them both: what should such an ordering criterion be, and why is any amount of risk negligible?

Where do these considerations leave us, with respect to 'the argument from decision theory'? *First*, they indicate that one need not assume that certain low probabilities are zero, if he wishes to understand why it is reasonable not to pay any finite amount for an infinite mathematical expectation. One could just as well assume, for example, either that it was not in his interest to pay for a game where $n = \infty$, or that, because of diminishing marginal utility, he could not do so. *Second*, as the example of Gibbon's illustrates, when the argument from decision theory is used to support a claim about *social policy*, it is either false (since $\epsilon \neq 0$) or question-begging, since it makes many presuppositions about what is acceptable.

2.3. *The Argument from Ontology*

A second method of attack, used by advocates of the 'probability-threshold' view, is 'the argument from ontology'. Proponents of this argument, made by nearly all assessors, subscribe to at least three claims. They maintain (Section 2.3.1) that the value of *societal* risk avoidance (also called its 'cost'), regardless of the type of risk, is a linear function of the annual probability of fatality, or a linear function of the actual probability, per unit time, of a unit cost burden occurring.[56] As Sagan puts it, Starr, Lave, and others assume "that we, the public, view all risks to life, e.g., death from an airplane accident, a risk of cancer, toxic chemical or biological exposure, in the same way, and that we would expend the same energies or resources to prevent [the same number of deaths from] each of these. Such an assumption is implicit in a comparison of qualitatively different risks which are quantitatively equal."[57] As a consequence of their subscribing to (2.3.1), proponents of the argument from ontology claim (2.3.2) that, for risks measurable in terms of the same unit cost burden (which is normally taken to be annual fatalities), those with greater probabilities ought to be attended to before smaller risks.[58] Therefore, say assessors, (2.3.3) because of the large number of risks whose annual probabilities of fatalities lie in the range from 10^{-2} to 10^{-5}, those whose probability is 10^{-6} or less ought to be ignored, because these risks are "negligible" or "insignificant".[59]

There are numerous problems with the first two premises (2.3.1) and (2.3.2) of the argument from ontology. I call these premises the linearity assumption and the commensurability assumption. I won't rehash them here, since I have discussed them elsewhere.[60] An assumption common to both these claims, however, is the heart of the argument that risks with certain low probabilities ought to be ignored, at least for the time being. This assumption is that the value of the risk avoidance is independent of any other variable except probability of a unit cost. If this assumption, which ontologically defines *value* in terms of *probability* of cost, is false, then there is strong reason to doubt that a given risk may be ignored as negligible, merely because its probability is quite low.

Perhaps the strongest reason for believing that the value of risk avoidance ought not to be defined purely in terms of probability is that such a definition ignores the importance of consequences. An extremely low probability of killing one person, for example, might be negligible. It is not clear, however, that if the consequences differed, then the risk would still be negligible. If this same low probability were that 150,000 persons might be killed, then the risk might not be said to be insignificant. This is why the probability-threshold position runs into such difficulties when used to evaluate risks such as those from commercial, nuclear-reactor accidents. Granted, the per-year, per-reactor probability of a nuclear core melt might be only 6×10^{-5}, an extremely low probability.[61] But if such an accident, according to the U.S. Brookhaven Report, could kill 145,000 people, cause $17 billion in property damage alone, and render an area the size of Pennsylvania uninhabitable,[62] then such a risk is hardly negligible. Perhaps, as Derek Parfit argues, a large number of fatalities "cancels out the smallness of the chance" and, for this reason, "it is not plausible to claim that a very tiny chance is no chance at all."[63]

The most basic argument against ignoring certain low probabilities is that their magnitudes, alone, are not sufficient grounds for determining their acceptability. If a risk is undesirable because it would affect great numbers of persons or because it is uncompensated or unfairly distributed, then just because it is small does not mean that it is morally negligible.

Consider the example of ocean burial of radioactive wastes, commonplace until twenty years ago. Using the probability-threshold position of Starr, Whipple, Okrent, Maxey, and others, one could easily argue that ocean burial of U.S. wastes imposes only a negligible increase in cancer risk on citizens of other countries whose shores are closest to the dumping grounds. If U.S. dumping is wrong because of its effects on citizens of other, innocent countries and because of our failure to compensate them for an admittedly small increase in health risks, then the smallness of the risk probability indeed mitigates the U.S. responsibility but does not eliminate it. Presumably, fairness demands that those who receive the benefits from a particular technology ought also to be the ones who bear its burdens, apart from the admitted smallness of many of those burdens.

Moreover, to justify ignoring societal evaluation of a given low-probability risk, purely because of its size, seems highly arbitrary. For one thing, it can easily be argued that such a position commits one to the naturalistic fallacy, to reducing ethics to science. This follows because a proponent of the probability-threshold view makes the assumption that a purely scientific property, having a low probability, is a sufficient condition for terming a risk acceptable.[64]

Second, if something is dismissed as acceptable or negligible, purely on the grounds of its low probability, then, as was mentioned earlier, the real question of its acceptability is begged. In arguing that a risk is negligible, one presumably is bound to tell why it should be accepted. This is because not all low-probability risks are worth the benefits derived from accepting them, particularly if questions of fairness, equity, or compensation arise. For example, most persons would probably agree that the health benefits derived from using computed-tomography, whole-body scans, as a way of annually screening the whole population for cancer, would not be worth even the low radiation risks involved.

Even if one were to subscribe to the thesis that the value of risk avoidance were independent of any parameter except probability of a unit cost occurring, there would still be at least two problems with assuming that, merely because certain probabilities were low, they were negligible. *First*, it is quite possible that, even if a given accident probability were extremely low, the aggregate of all relevant accident probabilities might be cumulatively very high. In this way, it is possible that, when take together, allegedly negligible probabilities of the large aggregate of known and unknown consequences, especially forgotten higher order-terms, could dominate the risk analysis.[65]

Second, regardless of the particular threshold used, from a practical standpoint, the actual risk for a given population is likely to exceed greatly the allegedly acceptable risk threshold. This is because certain subsets of persons (e.g., those with previous risk exposures to radiation, pesticides, etc.) bear a higher individual risk than others. As a consequence, acceptance of an allegedly safe *average* risk for public enterprises is almost certain to impose quite high risks on given *individuals* who are more sensitive than average to a particular exposure. Some of those who are more sensitive than average to toxic chemicals or to radiation include the elderly, children, those who lead sedentary lives, and victims of asthma. This raises the ethical issue of what it means to say that individuals have rights to equal protection. Does "equal protection" mean that all individuals have rights only to be protected against the same exposure levels of a particular toxin, or does it mean that each person has the right to be protected against the same degree of harm, regardless of the fact that quite different exposure levels could cause the same degree of harm in different people? As was already argued in Chapter Three (Section 3.1.1), with respect to a related claim (the commensurability presupposition), it is highly questionable to assume that use of a particular risk-exposure threshold, for all persons, is always ethically desirable. This is because sameness of exposure level does not guarantee sameness of risk protection; because sameness of protection is not the same as equality of protection; because sameness of concern or respect (in making the risk decision) is not the same as sameness of treatment; and because not all persons have equally justifiable claims to the same protection; rather, some individuals have defensible claims to better-than-the-same treatment owing to their rights to a reward for merit or virtue, or to their rights to recompense for past actions, or to the need for an incentive for future actions held to be highly socially desirable, or owing to their special needs. . . . But, rather than repeat the analyses of Chapter Three, suffice it to say that the same arguments employed there also can be used to show that acceptance of a threshold position presupposes acceptance of some highly questionable theses about equality.

A *third* problem with justifying the notion of risk thresholds by means of 'the argument from ontology' is the Starr–Whipple claim that the natural-hazards mortality rate (10^{-6} per year) presents a "lower limit for risk".[66] In assuming that this limit is a threshold below which risks are negligible, Starr, Whipple and others clearly presuppose that what is normal is moral, a highly doubtful assumption, and one repeatedly attacked by Moore.[67] The problem with assuming that a risk is moral, so long as its probability does not exceed that for 'normal' risks, is that it fails to answer the question of whether *any* increased risk is good and, if so, *why* it is good. Moreover, in assuming that an increase in risk, of the same magnitude as normal risks, is acceptable, one is really saying that it is acceptable to put certain numbers of lives in jeopardy, for whatever reason, so long as the total risk is statistically insignificant or comparable to normal rates of risk. In other words, one is assuming that a certain number of 'statistical casualties' are acceptable, even though it is widely acknowledged that the same number of predictable deaths of identifiable persons very likely would not be said to be an acceptable consequence of a given risk. The obvious question is whether one should accept risks known to cause some 'statistical casualties', especially if he would not accept the same risks, known to cause the same number of deaths of identifiable victims. This point, of course, recalls Gibbon's reply to Buffon. Would one be easy if he were one of the potential victims?

Perhaps the biggest problem, with assuming that what is normal is moral, is that one thereby commits himself to the assumption that the *status quo*, whatever it is, is acceptable or moral. On the contrary, the *status quo* could be either desirable, or undesirable, depending on a number of circumstances. On the one hand, for example, if mortality rates for natural hazards were inflated because of a number of easily remedied problems, e.g., inadequate medical care at the scene of the disaster, then surely such rates ought not to be called 'acceptable'. On the other hand, however, if the same rates were deflated because of a number of advantages, e.g., adequate medical care at the scene of the disaster, then perhaps these rates might be judged tolerable.

In dispensing with questions of responsibility, intention, fairness, equity, and justifiability, and subscribing to the notion of a negligible risk threshold based on a purely quantitative criterion, probability of fatality, Starr, Whipple, Okrent, Maxey, and others appear to ignore the complexity of the *ethical* dimension of risk assessment and public policymaking. They also appear to rely too heavily on the assumption, in defining the value of risk avoidance in terms of probabilities, that all risk probabilities can be determined with accuracy. For a number of reasons, they cannot. *First*, there are inaccuracies caused by modelling and measurement errors or assumptions. *Second*, there are events simply not amenable to analytical techniques, e.g., calculating the probability of human error or foul play. The U.S. loss of 3 astronauts on the ground, several years ago at Cape Kennedy, for example, vividly illustrates that even the best "no-cost-barred, no-cost-spared, systems analytic approaches in the world cannot anticipate every possibility",[68] especially where human error is involved.

Likewise, consider the difficulty of calculating the probability that some human foul play, such as cheating, lying, or sabotage, will occur, and thus change the allegedly objective probability of annual fatalities occurring from a given enterprise. For example, calculations reveal that the probability of a royal flush is 1 in 646,000. It could easily be argued, however, that the real probability of a royal flush occurring is much higher, especially if the probability of cheating is 1 in 10,000, which is several orders of magnitude higher.[69] Likewise one might also argue that a given catastrophe, whose probability was calculated to be 1 in 17,000 per installation, per year, has an actual probability that is much higher, in part because the calculated probability, typically employed in risk assessment, took no account of factors such as sabotage, terrorism, and human error.[70]

Third, with most events said to have catastrophic potential, but a low probability of causing annual fatalities, e.g., nuclear core melt, often there has not been enough experience with the technology to draw accurate empirical conclusions about either the likelihood or the consequences of all relevant risks, Moreover, a given accident frequency is compatible with a wide range of probability values. Consider one recent and interesting range of probabilities, that for the likelihood of a Three-Mile-Island type accident, as computed by Norman Rasmussen, author of the famous WASH-1400. That probability, according to Rasmussen, has a range from 1 in 250 reactor years to 1 in 25,000 reactor years.[71] Given typical probability estimates which vary by two orders of magnitude, it is difficult to feel confident that allegedly low probabilities (such as 10^{-6}) are accurate. Because of the inaccuracy of current nuclear-accident probabilities, nuclear opponents could be said to be correct in translating the lower-bound probability, for a population of 125 reactors, into 1 Three-Mile-Island-type accident, somewhere in the U.S., every other year. Both because of the probability *range* (with upper and lower limits differing by a factor of 100) used by experts such as Rasmussen, and because of the way the same probability can be translated in a particular context to sound better or worse than might first appear,[72] apart from any incorrect judgments, there is room for genuine disagreement, emphasis, or interpretation in viewing probabilities. For all these reasons, it is difficult to affirm with certainty that a given probability is actually below 10^{-6}.[73]

But if so, then it makes little sense to exclude certain allegedly low accident probabilities on the grounds that they are 'negligible' or impossible.

2.4. *Argument from Epistemology*

A third way in which assessors attempt to justify the probability-threshold view is by means of (what I call) 'the argument from epistemology'. Here they claim that, below a given level, usually 10^{-6}, persons are able neither to distinguish one small probability from another, nor to compare them. For this reason, they say, it is not meaningful to consider probabilities smaller than this threshold.[74] As two assessors put it, there is an "intuitively understandable" probability range, and below the lower limit of this range, probabilities cannot be known or compared meaningfully.[75] Hence, they claim, these probabilities are taken as zero and can be counted as negligible.

To illustrate their views, Starr and Whipple give the example of persons who choose to smoke one cigarette, or to take one automobile ride without buckling their seat belts. In both instances, the probability of fatality is very small. Starr and Whipple claim that these small risks are perceived as zero by those who take them, and that they are negligible.[76]

The obvious problems with counting subthreshold risk probabilities as negligible, because they are perceived as zero, are (1) that desirable risk policy is very likely a function of *actual* risks, not just *perceptions* of risks; and (2) that unknowable or indistinguishable risks are not necessarily negligible risks. The reasoning behind (1) is that persons can be harmed by what they don't know, and hence that they should be protected on the basis of what actual risks are and not merely on the basis of what they erroneously believe certain risks to be, especially when they underestimate those risks. The rationale behind (2) is similar; just because a risk is unknowable does not mean that it is negligible. To assume otherwise would be to commit the fallacy of the argument from ignorance. For example, I may not be able to measure parts per billion of asbestos in human lungs. If, because of this measurement impossibility, I assume that only parts per *million* of asbestos present a nonnegligible health threat, then I commit this fallacy. My ignorance, or the limited nature of my scientific knowledge, does not justify my assuming that parts per *billion* of this material are negligible. Apart from these two difficulties, (which come down to the simplistic assumption that 'what you don't know won't hurt you'), the argument from epistemology also rests on the doubtful presupposition that, because certain low probabilities are in fact perceived as zero, they are negligible.

Contrary to these presuppositions, however, one need not assume, either that very low probabilities are perceived as zero, or that only such a zero perception could explain why persons do not avoid certain minimal risks. In the case of the Starr–Whipple claim, for example, the smokers and the unbelted riders need not assume that their low-probability risk is zero. Rather, and more likely, as was already shown in discussion of the argument from ontology, they need only assume that the small risk of one cigarette or one unbelted ride is far outweighed by the benefit (e.g., pleasure, comfort) gained from either activity. In other words, the failure of some people to exhibit risk aversion regarding either smoking, or the failure to use seatbelts, can be explained in terms of factors other than minimizing an already low probability. This means that assessors who use the argument from epistemology, like those who employ the argument from ontology, beg the question of the centrality of probability regarding the value of risk avoidance; they do not explain why the value of risk avoidance is not a function of *trade-off* rather than merely a function of the *probability* of risk. To argue either that persons do, or ought to, make risk decisions, purely on the basis of either perceived or actual probability of fatality, is questionable because it presupposes that people are only concerned about physical threats to health and safety, rather than also about the trade-off among ethical, political, legal, and economic costs and benefits. Obviously a higher-probability risk might be more desirable than a lower-probability one, if the former risk were associated with an essential activity (e.g., use of given pesticides for disease control) whereas the latter were associated with a nonessential activity (e.g., use of given pesticides to enhance the cosmetic appeal of vegetables).[77] If this is so, then both the argument from ontology and the argument from epistemology fail because of the questionableness of their common assumption as to the sufficiency of probability magnitudes for defining acceptable risks.

3. CONCLUSION

Admittedly I have not examined the probability-threshold position from the point of view of its application to a great variety of contemporary technologies. My analysis has focused in large part on how this methodological stance has been exemplified in the case of nuclear fission; this is because it is the one technology toward which U.S. policymakers have clearly adopted a precise (10^{-6} average annual probability of fatality) probability-threshold position (see Section 2 earlier in this chapter). Following on the nuclear path, numerous risk assessors have urged that similar policy positions be adopted for other technologies (see the first two sections of this chapter). Using the nuclear-fission case is thus useful, not only because it provides the premier instance of the use of the probability-threshold position, but also because the same sorts of arguments which count for or against acceptance of this position in the nuclear case also count for or against it in other cases.

Moreover, although I have not investigated all the arguments used to support the probability-threshold position, some preliminary hypotheses might be proposed on the basis of this discussion of the arguments from decision theory, ontology, and epistemology. *First*, if all three arguments err in assuming that the magnitude of p alone provides a sufficient condition for judging the acceptability of given risks, then it may well be that, contrary to a 1977 NAS suggestion,[78] methods incorporating the probability-threshold position are inferior to those which do not, since the former, but not the latter, ignore the question of risk trade-off. *Second*, if all three arguments err in failing to consider how factors such as compensation and equity of risk distribution, affect the acceptability of risks below a certain threshold, then Kneese and others may well be correct in suggesting that more accurate risk assessment can be achieved by weighting the various risk-cost-benefit parameters according to alternative ethical criteria.[79] This, in turn, suggests that some form of adversary system is needed, in order to facilitate a choice among different, ethically weighted, risk assessments.[80]

Regardless of how the problems implicit in the probability-threshold position are handled, however, risk assessment ought not to proceed as it has in the past. All risk questions are ultimately philosophical questions. To attempt to reduce them to purely scientific issues, as do proponents of this position, is to ignore the value dimension of policy analysis and to disenfranchise the public who, in a democracy, ought to control that policy.

NOTES

1 A. Wildavsky, 'No Risk Is the Highest Risk of All', *American Scientist* **67** (1), (1979), 32; hereafter cited as: No Risk.

2 Wildavsky, No Risk, p. 33.

3 See Section 3.3.2 of Chapter Two earlier in this volume.

4 John Gofman and Arthur Tamplin, *Population Control Through Nuclear Pollution*, Nelson-Hall, Chicago, 1970, pp. 49–51.

[5] S. Gage, 'Risk Assessment in Governmental Decision Making', in *Symposium/Workshop . . . Risk Assessment and Governmental Decision Making* (ed. by Mitre Corporation), The Mitre Corporation, McLean, Virginia, p. 10; hereafter cited as: Gage, Risk, and Mitre, *RA*.

[6] J. Highland, 'Panel: Use of Risk Assessment . . .', in Mitre, *RA*, p. 632.

[7] A. Kneese, S. Ben-David, and W. Schulze, 'A Study of the Ethical Foundations of Benefit–Cost Analysis Techniques', working paper done under National Science Foundation funding, Ethics and Values in Science and Technology (EVIST) Program, 1979.

[8] See K. S. Shrader-Frechette, *Science Policy, Ethics, and Economic Methodology*, Reidel, Boston, 1984, Chapter 8; hereafter cited as: *Science Policy*.

[9] C. Zracket, 'Opening Remarks', in Mitre, *RA*, p. 3; hereafter cited as: Remarks.

[10] C. Comar, 'Risk: A Pragmatic *De Minimis* Approach', *Science* **203** (4378), (1979), p. 319; hereafter cited as: Risk. J. Hushon, 'Plenary Session Report', in Mitre, *RA*, p. 748; D. Okrent, 'Panel: Use of Risk Assessment', in Mitre, *RA*, p. 593; D. Okrent, 'Comment on Societal Risk', *Science* **208** (4442), (1980), 372–375; hereafter cited as: Comment. S. Gibson, 'The Use of Quantitative Risk Criteria in Hazard Analysis', in *Risk–Benefit Methodology* (ed. D. Okrent), UCLA School of Engineering and Applied Science, Los Angeles, 1975, p. 592; hereafter cited as: Quantitative. See W. Rowe, *An Anatomy of Risk*, Wiley, New York, 1977, p. 320; hereafter cited as: *AR*. See also H. Jellinek, 'Discussion', in Mitre, *RA*, p. 69, and Nicholas Rescher, *Risk*, University Press of America, Washington, D.C., p. 37; hereafter cited as: *Risk*.

[11] This thesis is that risks below a certain probability may be ignored. It is not to be confused with another type of threshold hypothesis, viz., that below a certain level of ingestion/exposure, a given substance (e.g., pesticide) is not harmful. The first claim is about the desirability of ignoring a certain level of known risk. The second claim, with which we are not concerned here, is about the level at which a known risk begins. For discussion of the latter claim, see K. S. Shrader-Frechette, *Environmental Ethics*, Boxwood, Pacific Grove, California, 1981, pp. 294–301.

[12] See note 10 preceding, as well as C. Starr, 'Benefit–Cost Studies in Sociotechnical Systems', in *Perspectives on Benefit–Risk Decision Making* (ed. by Committee on Public Engineering Policy, National Academy of Engineering, Washington, D.C., 1972; D. Okrent and C. Whipple, *Approach to Societal Risk Acceptance Criteria and Risk Management*, PB-271 264, U.S. Department of Commerce, Washington, D.C., 1977 (hereafter cited as: *ASRAC*); Hull, 'Discussion', in Mitre, *RA*, pp. 171–172; and Rescher, *Risk*, pp. 35–40.

[13] Okrent, Comment, p. 372.

[14] Okrent, Comment, p. 375.

[15] Gibson, Quantitative, p. 592.

[16] C. Starr, *Current Issues in Energy*, Pergamon, New York, 1979, p. 14; hereafter cited as: *CIE*. C. Starr and C. Whipple, 'Risk of Risk Decisions', *Science* **208** (4448), (1980), 1119; hereafter cited as: Risks. C. Starr, R. Rudman, and C. Whipple, 'Philosophical Basis for Risk Analysis', *Annual Review of Energy* **1** (1976), 630; hereafter cited as: Philosophical.

[17] U.S. Nuclear Regulatory Commission, *Reactor Safety Study: An Assessment of Accident Risks in U.S. Commercial Nuclear Power Plants*, WASH-1400, U.S. Government Printing Office, Washington, D.C., 1975, p. 38; hereafter cited as: WASH-1400.

[18] U.S. NRC, WASH-1400, p. 39. Despite their appeals for evaluating risks solely on the basis of their probabilities, the authors of WASH-1400 deny that they have addressed "the question of what level of risk from nuclear accidents should be acceptable by society" (U.S. NRC, WASH-1400, p. 248). Such a disclaimer is curious, however, given their repeated use of evaluative terms to describe nuclear risks. Terms like "negligible", "acceptable", and "insignificant", especially when combined with appeals to compare risks *solely* on the basis of probabilities, all give the reader the impression that the authors are promoting a number of value-laden premises regarding nuclear power. Commenting on WASH-1400, the U.S. Environmental Protection Agency (EPA) said as much in its remarks, which appear in Appendix XI of the document. This agency warns: "the comparative risk approach presented in the summary and in the main volume of the draft report is likely to imply an acceptability judgment to the average reader" (U.S. NRC, WASH-1400, Appendix XI, p. 2–2; see also L. Sagan, 'Public Health Aspects of Energy Systems', in *Energy and the Environment* (ed. by H. Ashley, R. Rudman, and C. Whipple), Pergamon, New York, 1976, p. 88; hereafter cited as: Public and *EAE*.) Even were an acceptability judgment not implicit, the repeated evaluation of the nuclear risk, purely in terms of probabilities, too easily lends itself to misuse by those who fail to understand the complexity of parameters whose consideration is necessary for judgments of risk acceptability. As the Union of Concerned Scientists pointed out, WASH-1400 all too easily lends itself to "being misused" (U.S. NRC, WASH-1400, Appendix XI, p. 2–14).

[19] U.S. NRC, WASH-1400, p. 226.

[20] U.S. NRC, WASH-1400, p. 247.

[21] Rescher, *Risk*, p. 36.

[22] See, for example, A. Wildavsky, No Risk, pp. 32–35. See Y. Aharoni, *The No-Risk Society*, Chatham House, Chatham, N.J., 1981, pp. 186–190; hereafter cited as: *NRS*. Finally, see Rescher, *Risk*, pp. 134–141.

[23] Wildavsky, No Risk, p. 33. See also Aharoni, *NRS*, pp. 187–189.

[24] Wildavsky, No Risk, p. 32. See also Aharoni, *NRS*, p. 53. See also Rescher, *Risk*, pp. 120–133.

25 For those who make this claim, see M. Maxey, 'Managing Low-Level Radioactive Wastes', in *Low-Level Radioactive Waste Management* (ed. by J. Watson), Health Physics Society, Williamsburg, Virginia, pp. 410, 417; see also Wildavsky, No Risk, p. 37.

26 Wildavsky, No Risk, p. 32.

27 Wildavsky, No Risk, p. 34.

28 B. Cohen and I. Lee, 'A Catalog of Risks', *Health Physics* **36** (6), (1979), 707–722; hereafter cited as: Catalog. Wildavsky, No Risk, p. 33; see also N. Rescher, *Unpopular Essays on Technological Progress*, University of Pittsburgh Press, 1980, pp. 45–48; hereafter cited as: *UE.*

29 Wildavsky, No Risk, pp. 36–37. Aharoni, *NRS*, pp. 177–180.

30 Starr and Whipple, Risks, p. 1119. See Aharoni, *NRS*, esp. p. 177.

31 Allen Buchanan, 'Revolutionary Motivation and Rationality', *Philosophy and Public Affairs* **9** (1), (Fall 1979), 64–66; hereafter cited as: Motivation.

32 Frank Miller and Rolf Sartorius, 'Population Policy and Public Goods', *Philosophy and Public Affairs* **8** (2), (Winter 1979), 158–160; hereafter cited as: Policy.

33 Mancur Olson, *The Logic of Collective Action*, Harvard University Press, Cambridge, 1971, p. 64.

34 Rescher, *Risk*, p. 36.

35 L. Savage, *The Foundations of Statistics*, Wiley, New York, 1954, pp. 91 ff; hereafter cited as: *Foundations.*

36 K. Arrow, *Essays in the Theory of Risk Bearing*, Markham, Chicago, 1971, p. 14; hereafter cited as: *ETRB.*

37 Arrow, *ETRB*, p. 5.

38 Arrow, *ETRB*, p. 5.

39 Arrow, *ETRB*, p. 14.

40 Savage, *Foundations*, p. 94.

41 Savage, *Foundations*, pp. 95–97.

42 Arrow, *ETRB*, p. 14, makes a similar point.

43 J. Keynes, *Treatise on Probability*, Macmillan, London, 1929, p. 322; hereafter cited as: *TOP.*

44 Arrow, *ETRB*, p. 15.

45 Brian Barry, *Sociologists, Economists, and Democracy*, Collier-Macmillan, London, 1970, p. 32.

46 Keynes, *TOP*, p. 322.

47 This argument is made by Rescher, *Risk*, p. 39.

48 Arrow, *ETRB*, p. 22 ff.

49 Arrow, *ETRB*, pp. 23–24.

50 Arrow, *ETRB*, p. 24.

51 Arrow, *ETRB*, p. 25.

52 Arrow, *ETRB*, pp. 26–28.

[53] Arrow, *ETRB*, p. 28.
[54] Arrow, *ETRB*, pp. 22–28.
[55] Arrow, *ETRB*, p. 28.
[56] See, for example, Starr, *CIE*, p. 23; Starr and Whipple, Risks, pp. 1115–1116; Starr, Rudman, and Whipple, Philosophical, pp. 636–637; Cohen and Lee, Catalog, p. 707; and L. Philipson, 'Panel on Accident Risk Assessment', in Mitre, *RA*, p. 385.
[57] L. Sagan, 'Public Health Aspects of Energy Systems', in Mitre, *RA*, p. 88.
[58] Comar, Risk, p. 319; Starr, *CIE*, p. 12; Gibson, Quantitative, p. 599.
[59] See, for example, Okrent, Comment, p. 375; Starr and Whipple, Risks, p. 1119.
[60] See Chapters Three and Six of this volume for discussion of these assumptions.
[61] This probability is generally accepted in the U.S. nuclear industry. It is given in: U.S. Nuclear Regulatory Commission, Reactor Safety Study – *An Assessment of Accident Risks in U.S. Commercial Nuclear Power Plants*, Report No. (NUREG-75/014) WASH-1400, Government Printing Office, Washington, D.C., 1975, pp. 157 ff; hereafter cited as: WASH-1400. Note, however, that when this probability is applied to all nuclear plants now under construction or in operation, the lifetime probability of a core melt is 1 in 4. (For these calculations, see K. Shrader-Frechette, *Nuclear Power and Public Policy*, second edition, Reidel, Boston 1983, pp. 84–85; hereafter cited as: *Nuclear Power*.
[62] These statistics are given in U.S. Atomic Energy Commission, 'Theoretical Possibilities and Consequences of Major Accidents in Large Nuclear Power Plants', U.S. AEC Report WASH-740, Government Printing Office, Washington, D.C., 1957, and its update, i.e., R. J. Mulvihill, D. R. Arnold, C. E. Bloomquist, and B. Epstein, 'Analysis of United States Power Reactor Accident Probability', PRC R-695, Planning Research Corporation, Los Angeles, 1965; hereafter cited as: WASH-740-UPDATE.
[63] Derek Parfit, 'Correspondence', *Philosophy and Public Affairs* **10** (2), (Spring 1981), 180–181. See also Joel Yellin, 'Judicial Review and Nuclear Power', *George Washington Law Review* **45** (5), (1977), pp. 982–983, who makes a similar point.
[64] For information on the naturalistic fallacy, see G. E. Moore, *Principia Ethica*, University Press, Cambridge, 1951, pp. viii–ix, 23–24, 39–40, 73, 108; hereafter cited as: *PE*. See also Shrader-Frechette, *Nuclear Power*, pp. 136–168.
[65] A. Lovins, 'Cost–Risk–Benefit Assessment in Energy Policy', *George Washington Law Review* **45** (5), (1977), p. 934, makes this same point.
[66] Starr and Whipple, Risks, p. 1119; Starr, *CIE*, p. 15.
[67] Moore, *PE*, pp. 43, 58.

[68] H. Hollister, 'The DOE's Approach to Risk Assessment', in Mitre, *RA*, p. 50. See also S. Samuels, 'Panel: Accident Risk Assessment', in Mitre, *RA*, p. 415.

[69] W. Fairley, 'Criteria for Evaluating the 'Small' Probability', in *Risk–Benefit Methodology* (ed. D. Okrent), UCLA School of Engineering and Applied Science, Los Angeles, 1975, pp. 406–407.

[70] This is exactly the case with the risk of nuclear core melt. See note 61 earlier.

[71] N. C. Rasmussen, 'Methods of Hazard Analysis and Nuclear Safety Engineering', in *The Three Mile Island Nuclear Accident* (ed. by T. Moss and D. Sills), New York Academy of Sciences, New York, 1981.

[72] A similar example of the context-dependent character of nuclear probabilities is given in Shrader-Frechette, *Nuclear Power*, pp. 84–85. Here the author shows how the same core melt probability, 1 in 17,000 reactor years can also be translated as 1 in 4, given 150 reactors operating for 30 years. See note 61.

[73] See T. Feagan, 'Panel: Human Health Risk Assessment', in Mitre, *RA*, p. 291, and L. Philipson, 'Panel on Accident Risk Assessment', in Mitre, *RA*, p. 476.

[74] See, for example, F. Farmer, 'Panel: Accident Risk Assessment', in Mitre, *RA*, pp. 396–397.

[75] C. Starr and C. Whipple, Risks, p. 1117.

[76] C. Starr and C. Whipple, Risks, p. 1117; Starr, *CIE*, p. 18.

[77] In an earlier work, Okrent and several other assessors proposed a much more defensible version of the threshold hypothesis. They argued that technological risks could be classified as essential, beneficial, or peripheral to society, and that the maximum acceptable risk to the individual was different, depending on the risk classification. Risks in the first class were acceptable if they posed no higher than a 10^{-4} annual probability of death. Those in the second class, if they posed no greater than a 10^{-5} probability, and those in the third class, if they posed no more than a 10^{-6} probability. The chief merit of this earlier formulation is that it appears to take some account of the reasons why a particular threshold might be acceptable. If does not appear to rest on the implausible assumption of the later article, viz., that probability (or magnitude) alone is a sufficient condition for determining whether a risk is negligible. (See Okrent and Whipple *ASRAC*, pp. 8, 19, 20.)

[78] Zracket, Remarks, p. 3.

[79] See endnotes 7 and 8 above.

[80] Shrader-Frechette, *Science Policy*, Chapters 8–9, discusses this point.

CHAPTER SIX

THE LINEARITY ASSUMPTION

1. INTRODUCTION

In the U.S. today, more than ten federal statutes contain statements about risk assessment. The process of setting environmental standards, for anything from workplace exposure to benzene, to consumer exposure to acrylonitrile in beverage containers, increasingly rests on the analysis of risk. The U.S. courts have made multiple references to risk assessment in reviewing the actions of regulatory agencies. Yet, both the criteria for a successful risk assessment and its role in helping to set policy are subjects of controversy among experts as well as the public. One reason for such disputes is that risk *estimation* often must rely on controversial methodological presuppositions, while risk *evaluation* often must appeal to debatable ethical and policy assumptions.[1]

As we have become aware of the finite nature of our resources, U.S. agencies, courts, and citizens have begun to reject risk policies based on eradicating hazards to the greatest degree possible. The costs of environmental regulation appear to be perceived as more and more burdensome. As a consequence, government, industry, scientists, and society are demanding a new criterion for the value of risk abatement.

As was argued in the previous chapter, one of the typical criteria for risk abatement is the probability-threshold position, the belief that risks below a certain annual probability of fatality ought to be ignored. As was also argued in this chapter, risk assessors make a number of doubtful logical, ethical, and epistemological assumptions whenever they subscribe to the methodological tenets which

I have called "the probability-threshold position." Underlying adherence to this commonly accepted policy position is a more basic belief that, because a particular risk probability is small, the value of avoiding the risk is correspondingly small. In other words, assessors are able to conclude that a certain probability of fatality justifies neglecting abatement of the risk with which this probability is associated only because of their prior adherence to two other beliefs. These are that probability of fatality determines the degree of risk acceptability and that risk acceptability and probability of fatality are inversely related. (These latter two beliefs comprise what I call "the linearity assumption".) This means that the linearity assumption and the probability-threshold position are closely related; the former is perhaps the most important necessary condition upon which the latter rests. But if so, then a complete analysis of the probability-threshold position requires that one evaluate the linearity assumption as well. Moreover, as we shall soon see, the linearity assumption is important in its own right, not only because the risk literature contains no analytical, philosophical evaluations of it, but also because a number of members of the risk-assessment community use it as one of the foundations for their work.

Admittedly, because there is no general agreement among risk experts as to the necessary and sufficient conditions for a successful assessment, it is highly unlikely that acceptance of the linearity assumption is universal among all members of the risk-analysis community. Cohen, Comar, Gibson, Lee, Maxey, Okrent, Rasmussen, Rudman, Starr, and Whipple, however, to name but a few prominent practitioners of the method of revealed preferences, all subscribe to the linearity assumption. Because of their adherence, and because of the dramatic policy consequences of their adherence (see the last section of this chapter), it is important to examine the philosophical underpinnings of this assumption.

Succinctly put, the linearity assumption is the belief that there is a linear relation between the actual *probability* of fatality associated with a given risk and the *value of avoiding that risk*. As should be immediately obvious, the criterion has great intuitive appeal. It is reasonable to assume that the value of society's avoiding a certain risk grows in proportion as the probability of harm increases. For example, if the carcinogenic potential of a particular chemical were extremely small, then the value of avoiding it would likely also be small. If its cancer-causing potential increased, however, then one might expect the value of avoiding the chemical to increase to a similar degree. In fact, many assessors believe that this assumption provides a sufficient explanation of observed societal behavior regarding risks having different probabilities of causing fatalities. Admittedly, however, alleged past 'choices' of society may have been imposed, not chosen, and admittedly, past behavior regarding risk acceptance provides a sufficient criterion neither for what ought to have been done in the *past*, nor for what ought to be done in the *future*. Nevertheless, many assessors maintain that the probability of fatality of accepted past and present societal risks provides a sufficient criterion for judging the acceptability of future social risks.[2]

2. THE FATALITY INTERPRETATION OF THE 'LINEARITY ASSUMPTION'

As Starr, Rudman, and Whipple formulate this assumption which they and many other risk assessors hold, there is "a linear relation between [societal] risk and the cost of that risk." They define 'risk' as "the actual *probability* per unit time of a unit cost burden occurring", where unit cost burden is typically taken to be a given number of annual fatalities, usually one. Likewise, they define 'cost' "in terms of injuries (fatalities or days of disability) or other damage penalties (expenses incurred) or total social costs (including environmental intangibles) of that risk".[3] Thus defined,

I have called this tenet 'the linearity assumption'. In practice, many (if not most) assessors (employing this assumption) interpret the "probability per unit time of a unit cost burden occurring" to be "annual probability of [one] fatality",[4] This I call the 'fatality interpretation'. Hence the fatality interpretation of the linearity assumption is that there is a linear relation between the actual "probability of fatality" and "the value of risk avoidance" or the "cost of a risk".[5]

Comar and others claim, for example, that to "avoid squandering resources" we must not "reduce small risks while leaving larger ones unattended."[6] They say, implicitly, that risks with greater probability ought to be attended to before smaller ones. Likewise Okrent claims that "in view of their statistically smaller contribution to societal risk, major accidents may be receiving proportionately too much emphasis compared to other sources of risk."[7] In making this suggestion, he too appears to propose that something should be viewed as hazardous solely on the basis of its likelihood. Gibson makes this point even more directly when he claims that there is no reason to distinguish single- versus multiple-fatality accidents, so long as the risk is of a certain level.[8] In fact, says Rasmussen, viewing risks in this manner is a matter of "consistency in public attitudes".[9] This suggests that anyone who does not value risks in such a simple manner may be accused of inconsistency.[10]

By adhering to the fatality interpretation of the linearity assumption, many assessors subscribe to at least three doubtful tenets: (1) that a unit cost is adequately represented by *fatalities*: (2) that the value of risk avoidance is *independent* of any variable (e.g., benefits), except probability of fatality; and (3) that there is a *linear relation* between societal risk (expressed in terms of probabilities) and the value of risk avoidance. Since (1) and (2) seem to me to be easily shown to be implausible, and since (3) has (to my knowledge) not been adequately assessed in the risk analysis literature, I wish to focus on this last claim. Specifically, I want to discuss one argument against tenet (3), viz., that it is highly implausible, since people often have good reasons for viewing risks of fatality in nonlinear ways. That is, they frequently have good reasons for placing higher value on risk avoidance for events having a low probability of fatality than for those having a higher probability. For example, the public values airplane safety more than automobile safety.

Assessors, however, maintain that safety policy should be made on the basis of the thesis that risks with a higher probability of fatality are less acceptable than those with a lower value. Assuming that consistent and objective safety policy must address risks whose probability is greater before those whose probability is less, Cohen and Lee, for example, rank 54 health risks solely according to their decreasing order of probability. They claim that the ordering in their table "should be society's order of priorities" for health and safety programs.[11]

Many assessors argue, likewise, that it is "inconsistent" for the public both to tolerate 50,000 automobile deaths per year and yet to be alarmed at generating electricity through nuclear fission.[12] Maxey accuses those who view nuclear risks as more dangerous than other, more probable, risks, of having "pathologic fear" and "near clinical paranoia". Numerous authors, including Starr, Whipple, Okrent, Maxey, Cohen, and Lee, maintain that if the public only understood the *probabilities* (of fatalities) of the risks involved, they would not fear statistically less significant hazards, like LNG (liquefied natural gas) or nuclear accidents, more than some more likely ones.[13] In other words, since they assume that there ought to be a linear relationship between the actual probability of a fatality, and the value of risk avoidance, many assessors do not view the observed societal aversion to certain low-probability risks as evidence against the fatality interpretaion of the linearity assumption.[14]

Assessors explain such counterexamples (e.g., aversion to nuclear fission) as simply a result of the fact that the public does not know the accident probabilities in question. In so doing, they assume that no good arguments can be brought against the fatality interpretation. Their assumption is likely false, however. For one thing, the restriction of risk to probability of fatality is highly questionable, since there are obviously many other 'cost burdens', e.g., 'decreasing the GNP by a given amount', whose probability also determines the value of avoiding a given risk. Another problem is that the fatality interpretation fails to consider the facts that benefits play a role in risk avoidance, and that the value of avoiding a given risk is often a function of the benefits to be gained from taking the risk. The value of risk avoidance is rarely simply a function of the probability of fatality. Many other factors also affect how it is valued. For example, a risk that is otherwise acceptable may well be unacceptable if it is imposed involuntarily or distributed inequitably.[15] In fact, if Fischhoff and other assessors who employ psychometric surveys are correct, risk acceptability is more closely correlated with equity than with other factors, including voluntariness.[16]

Likewise, even though they have the same expected value, there is reason to believe that one accident killing 10,000 people is worse than 10,000 accidents, each killing one person. For one thing, society probably could not recover from 4 billion simultaneous deaths, even if the accident causing them occurred only once in 10,000 years. That sort of event is clearly worse than the preventable deaths caused by cigarette smoking, even though the cigarette-induced deaths "occur at the same average rate".[17]

Another factor that might account for high aversion to low-probability events is their catastrophic potential, or the fact that low-probability/high-consequence situations are often the product of public, as opposed to private, risk. Since there is evidence that the psychological trauma (feelings of impotence, depression, rage) associated with the imposition of a public risk is greater than that associated with the choice of a private risk of the same probability,[18] there is good reason for society's risk aversive behavior (with respect to potentially catastrophic, low-probability events) not to be proportional to the probability of fatality. Moreover, although according to utility theory, a high-probability/low-consequence event and a low-probability/high-consequence situation may have the same expected value, it is obvious that it is often reasonable to be more averse to the low-probability/high-consequence situation. For this reason, some theorists have conjectured that a large accident with an annual average of *n* fatalities might have an importance greater than that attributed to many accidents which together cause *n* fatalities.[19]

If it is true that an accident's importance is not measured simply by its probability and magnitude, but is affected by factors such as whether it is public, uncompensated, inequitably distributed, or catastrophic, then it makes sense for people to value the same level of safety differently in different settings. Even though this may be termed economically inefficient, it is neither inconsistent nor irrational.

The relative values of avoiding airplane and automobile accidents, for example, are rarely functions simply of the respective probabilities of airplane and automobile fatalities. If, for instance, one belongs to the class of automobile drivers who are non-smokers, teachers, between the ages of 25 and 65, who have not been ticketed for speeding and who have not been involved in an auto accident in the last three years, then the probability of being involved in an auto accident is considerably lower than that for a person not belonging to the class of people having these characteristics. For such a person, it might be eminently reasonable (not irrational and inconsistent) to be more averse to airplane travel and to view it as riskier for him than auto travel. It might also be reasonable for society to take account of one's personal contributions to auto safety, and of the different distributive effects of various safety programs, and therefore to decide to pay more for airplane safety than for auto safety. Considerations such as the chances of survival, given an accident, and the catastrophic nature of aviation accidents might also play a role in a societal decision to pay more for aviation safety.

Another reason, why the public might not be more averse to risks of allegedly greater magnitude or probability, has to do with the source of the funds expended on safety programs or equipment. If safety programs for airplanes are financed largely by the industry or by passengers, while safety equipment for automobiles is paid for by the owner, then the public might prefer to beef up standards for airplane safety prior to those for automobile safety.

Starr, Whipple, and others, however, must claim that auto safety ought to be valued more highly than plane safety because the per-mile risk probability is greater for auto, than for airplane, travel. Although they maintain that there may be a linear relationship between societal risk and cost, and that there is a continuous, straight-line trade-off between societal risk and cost,[20] it is not clear why they claim that the cost of a risk is linear with its probability. Such a simplistic relation seems highly unlikely.

A third criticism, the specific one on which I wish to focus, is that the fatality interpretation, and especially tenet (3), run counter to the way that people often evaluate risks, especially low-probability catastrophic risks. As Zeckhauser points out,[21] once a new element of risk is announced or imposed, it gives citizens something to think and worry about. But doesn't it seem unlikely that a minor risk, say 1/10 of 1%, would generate *only* 1/10 of the aversion or anxiety of a 1% risk of the same magnitude? Small probability risks of great possible magnitude, in fact, seem likely to be valued in a way that is *not* linear. If one admits that the anxiety cost is a substantial portion of the amount that one might pay to avoid a given risk, then the case of anxiety costs may well falsify the thesis of linearity.[22]

3. THE ARGUMENT THAT LOW PROBABILITY EVENTS ARE PERCEIVED AS LIKELY

To support their claim about the public's alleged ignorance of correct probabilities, assessors often appeal to the notion that, below a certain level (say 1 in 10^4), the public cannot distinguish one very small probability from another. Their argument is that there is an intuitively understandable probability range, within which the public's judgments are consistent with the fatality interpretation of the linearity assumption, and analytic methodology, but that, outside this range (where most of the counterexamples fall), the public "has not learned" to view hazards consistently, by means of this assumption, and instead relies on intuitive, subjective, and incorrect assessments of risks. Starr, for example, asserts that, "unlike the natural catastrophes – earthquakes, typhoons, floods, tidal waves, etc., – society has not learned to place such hypothetical man-made events [like nuclear catastrophes] in an acceptable comparative perspective, particularly when they are poorly understood by the public."[23] Generalizing on the basis of the nuclear power case, Starr criticizes public "concern with imaginary large catastrophes",[24] and suggests that controversies over technology arise "because of intuitive estimates of unreasonably high risk".[25]

According to Starr and Whipple, since low-probability catastrophies lie outside the range of intuitively understandable probabilities, their likelihood is misperceived by the public.[26] To support this thesis about the misperception of low-probability events, Starr and Whipple cite two arguments, one general and the other particular. They note, *in general*, that extremely unlikely occurrences (winning a lottery, for example) are perceived as quite probable. They also maintain, *in particular*, that in the paradigm case of low-probability, catastrophic risk, that of nuclear power, extremely unlikely accidents are viewed as quite probable.[27] Let us examine each of these arguments.

At least two difficulties face the first, or *general*, argument about misperception of probabilities, especially of catastrophes. First, not all low-probablity events are seen as being highly likely. The fact that the low-probability event of a meteor striking the earth is in fact perceived as a highly unlikely event is precisely why a number of nuclear proponents have used it in an analogy to explain the predicted frequency of a given nuclear accident.[28] Moreover, if some low-probability events are in fact perceived as being extremely unlikely, while others are not, then contrary to Starr's and Whipple's suggestion, the reason for misperceiving a certain low probability appears to be something other than merely whether it falls outside some allegedly understandable probability range.

A second difficulty with this general argument about misperception of probabilities is that not all cases of public aversion to low-probability, catastrophic risks can be attributed to the belief that the catastrophe has a higher probability of fatality. The *reason* for society's aversion to certain low-probability events could well be something other than that their likelihood is misperceived, because they fall outside some allegedly understandable probability range. It is quite possible that the actual probabilities of fatalities are *accurately* perceived, but that other factors, e.g., the low level of benefits arising from the risk, cause societal aversion to low-probability accidents. Likewise, it is quite possible that the actual probabilities of fatalities are in fact *misperceived*, but that the dearth of benefits, and not these misperceived probabilities, plays the dominant role in causing societal aversion. In fact, precisely this possibility appears to be the case with respect to public perception of nuclear-accident probabilities.

4. THE ARGUMENT THAT LOW NUCLEAR-ACCIDENT PROBABILITIES ARE PERCEIVED AS HIGH

To support their claim that the public is often ignorant of correct risk probabilities, many assessors maintain that laymen view low-probability nuclear accidents as quite likely. How valid is this particular belief? Let us evaluate the Starr–Whipple defense of it.

Citing the work of Otway, Lawless, Fischhoff, *et al.* to support their claim, Starr and Whipple argue that "the bulk of disagreement" over nuclear power is over different beliefs about accident probability.[29] Starr, for example, claims that, because laymen do not understand atomic energy, they erroneously believe that serious nuclear accidents are far more likely than natural catastrophes, such as floods.[30] Alleging that most nonscientists have a disproportionate fear of complex technologies, Starr criticizes the public's worry about (what he says are) highly unlikely, catastrophic, nuclear accidents.[31] He claims that societal controversy over commercial atomic energy arises solely because uninformed, fearful persons rely on incorrect, merely "intuitive", estimates of nuclear accident probabilities.[32]

In thus discussing the causes of controversy over technology, Starr, Whipple, and other assessors make at least three problematic assumptions. These are: (1) that the work of Otway, Lawless, Fischhoff, *et al.* shows that controversy over technology arises primarily becauses of incorrect perceptions of accident probabilities; (2) that one can generalize about controversy over technology on the basis of particular conclusions about controversy over nuclear fission; and (3) that "the bulk of disagreement over nuclear power" is over different beliefs about accident probability. Let us consider first assumption (1) regarding the work of Otway, Lawless, Fischhoff, and others.

Although Starr and Whipple assert that Otway's research indicates that "the bulk of disagreement" over nuclear power is over different beliefs about risk, about accident probability, Otway's published studies provide little or no support for the claim that Starr and Whipple attribute to him. In fact, there are at least four reasons why Otway's own remarks indicate that he believes that the bulk of disagreement is over *values*, and over *benefits* attributed to nuclear power, and not over different beliefs about accident probabilities. *First*, he says that "in general, the con [nuclear power] group . . . assign high importance to the risk items while the pro group view benefit-related attributes as most important."[33] This suggests that the disagreement is over the value accorded to risks and benefits, and not over the actual accident probabilities themselves.

Second, Otway claims that although both pro and con groups "strongly believe that nuclear power is in the hands of big government or business, . . . the pro group evaluates this attribute positively, the con group evaluates it negatively."[34] *Third*, says Otway, his results indicated that there are only three statistically significant differences between "the cognitive structures which most clearly differentiate between the two groups". These three differences all concern "the benefits of nuclear power". They are its ability to provide benefits essential to society, good economic value, and a higher quality of life. The pro group, he says, "strongly believed that nuclear power offers these benefits while the con group tended to be uncertain to somewhat negative."[35] Most important, *fourth*, Otway explicity states: "There were no significant differences between the [pro-nuclear and anti-nuclear] groups on the eb [evaluation-belief] scores of any items related to risk."[36] Thus his own statement appears to be a flat contradiction of the Starr-Whipple claim that Otway's research shows that the bulk of disagreement over nuclear power is over beliefs about accident probabilities.

Likewise, it is not clear at all that Lawless's work supports the thesis that controversy over technology arises largely because of erroneous risk estimates on the part of the public. In fact, Lawless does not mention that misperceived probabilities cause disagreements. On the contrary, he argues that the degree of controversy over technology is greatest where proof of given effects, as in low-level radiation, is most difficult to obtain.[37] In most cases of controversy over technology, he says, certain scientific evidence was necessary for decisionmaking, yet in 36 of the 45 cases he studied, the evidence was insufficient for decisionmaking.[38] In other words, controversy appears to be largely a function of the *uncertainty* in scientific knowledge, not of the incorrect public perception of *certain* knowledge. Lawless also argues that nuclear controversy, in particular, has been caused by the lack of credibility of the (Atomic Energy Commission) AEC,[39] and by apparent government failure to consider environmentalists' values.[40] He notes, in general, that controversy over technology arose, in more than 50% of the cases studied, because technologies were allowed to grow even after evidence of a problem had been observed, and because they were used irresponsibly.[41] All this suggests that controversy over technology is not unequivocally a product of erroneous risk perception by the public.

As with the Otway and Lawless studies, it also is not clear that the work of Fischhoff, *et al.* supports the thesis either that controversy in general arises because of misperception of accident probabilities, or that controversy over nuclear power arises because of the misperception. Their research with the League of Women Voters indicates that the nuclear risk was not perceived as worth the allegedly low benefits accruing from it.[42] If Fischhoff's conclusion is correct, then much of the controversy over nuclear power might be over the issue of trade-off, rather than over that of misperceived probabilities.[43] In any case, it provides plausible grounds for rejecting the interpretation, either that the value of risk avoidance is linear with the actual probability of fatality, or that misperceived nuclear probabilities cause this relation to be viewed in a nonlinear way.

Also contrary to Starr's position, Fischhoff, *et al.*, specifically note that, on their surveys, the public (students and members of the League of Women Voters) judged nuclear power to have "the *lowest* fatality estimate" for the 30 activities studied, but the "*highest* perceived risk".[44] As a consequence, they conclude, "we can reject the idea that laypeople wanted to equate risk with annual fatalities, but were inaccurate in doing so. Apparently, laypeople incorporate other considerations besides annual fatalities into their concept of risk."[45]

On the Fischhoff study, the key considerations influencing judgments of high risk were not perceptions of high accident probability but the facts that certain technologies, like nuclear, represented an unfamiliar, as opposed to a common, risk; an inequitably, as opposed to equitably, distributed risk; and that they posed severe consequences, in the unlikely event that an accident were to occur.[46] In fact, the assessors found that perceived risk could be predicted almost completely accurately solely on the basis of the single variable, "severity of consequences", even though the probability of those consequences occurring was quite small and was perceived as quite small.[47]

If this is true, then Starr's suggestion that controversy is fueled primarily by incorrect probability estimates is less helpful than the suggestion that, in cases of high-magnitude events, it is the possible consequences, not their perceived probability of occurrence, that is important to societal evaluation. Wilson apparently believes that this is the case, and proposes that N lives lost simultaneously in a catastrophic accident should be assessed as a loss of N^2 lives. In other words, Wilson argues that the risk conversion factor for catastrophic accidents should be N^2, as a function of N fatalities.[48]

Admittedly the studies by Fischhoff and others show only that, contrary to Starr, controversy over certain technologies very likely arises because of values placed on consequences, not because of overestimated risk probabilities. The studies do not show that consequences *ought* to be valued in this way. Moreover, the AEC

(Atomic Energy Commission), the NRC (Nuclear Regulatory Commission), and the courts generally "have consistently taken the position that probabilities are determinative of risk, regardless of potential consequences."[49] Nuclear risk assessments have also consistently adopted the nuisance rule that probabilities alone determine risk. The basis for such a rule has been society's interest in technological development. Historically the rule has owed its inspiration to "the reluctance of the 19th century courts to allow the traditionally restrictive law of nuisance to hinder economic progress."[50]

There are, however, a number of reasons for arguing that, in certain cases, the consequences are more important than the accident probabilities. First, although assessors, government agencies, and the courts often define risk simply in terms of probability, such a definition fails to account for the greater social disruption arising from one massive accident, as compared to the social disruption caused by many single-fatality accidents killing the same number of people.

Second, the law of torts recognizes the heightened importance of high-consequence events, apart from their probability of occurrence. It allows for application of the rule of strict liability for abnormally dangerous activities. Yellin affirms that, for the rule of strict liability in law, risk is based almost totally on grave potential consequences, regardless of the associated probability.[51] Part of the justification for this judicial emphasis on accident consequences is apparently the fact that the parties involved in litigation over catastrophic accidents, viz., the injured persons and the persons liable for the injury, are not equal in bargaining power. The representative of some technological or industrial interest usually has more clout than the person damaged by it. Moreover, a person is more deserving of compensation according to strict liability when he is victimized by an impact that he did not voluntarily accept or help to create, as is often the case with high-consequence events. Because of the inequivalence between parties in liability suits involving catastrophic technological accidents, it is more likely that laws sensitive to consequence magnitudes are needed to insure attention to serious public health effects and to provide limits to dangerous impacts not comparable to those in our previous experience.[52] For both these reasons, then, there appear to be plausible grounds for denying that controversy over technology is primarily controversy over accident probabilities and for affirming that societal risk evaluation of potentially catastrophic technologies ought to focus on the value of accident magnitudes, rather than only on their probabilities.

In addition to these general considerations, there are a number of other reasons for questioning Starr's claim that "the bulk of disagreement over nuclear power" is over probabilities. For one thing, such a claim seems both to underestimate the agreement of nuclear proponents and opponents on the facts and probabilities of the matter, and to overestimate the extent of their consensus on political, ethical, social, legal and economic assumptions relevant to the debate. As one attorney expressed it, "current debate over whether nuclear power is safe or unsafe emerges as a spurious issue, for both sides recognize that nuclear power is an inherently dangerous technology."[53]

Or, as one regulatory agency spokesperson put it, all those involved in the controversy "recognize the inherent danger of nuclear power plants."[54] If this is true, then perhaps the nuclear controversy may be viewed, not only as a conflict about degree of nuclear danger, i.e., about correct and incorrect risk probabilities, as Starr and Whipple appear to believe, but also as a controversy over whether the alleged danger is worth the benefits it brings.

Moreover to claim that "the bulk of the disagreement" over nuclear power has been caused by "intuitive estimates of unreasonably high risk" seems to presuppose that there is no room for reasonable disagreement over nuclear risk probabilities; Starr and Whipple appear to believe that it is "perfectly valid to base public policy on expert estimates and data", but that, once an expert has spoken, any disagreement with his views must be unreasonable and intuitive.[55] Such a notion is doubly implausible.

First, it presupposes a far more objective and coherent picture of nuclear risk data than is now available. Even the authors of the most complete study of nuclear risks, WASH-1400, themselves cautioned that their probability estimates were deficient, unprovable, possibly incomplete, assumption-laden, and saddled with "an appreciable uncertainty".[56] They clearly pointed out that "the present state of knowledge probably will not permit a complete analysis of low-probability accidents in nuclear plants with the precision that would be desirable."[57] In the face of such caveats, Starr's and Whipple's alleged certitude, about which nuclear risk probabilities are correct and which are incorrect, appears puzzling. Their assurance that the public overestimates nuclear-accident probabilities is also puzzling, both because of the Fischhoff findings cited earlier and because there is virtually no hard data upon which to base such assurance. Relatively new events, like nuclear accidents, have not generated enough empirical data to allow one to affirm a specific accident frequency with accuracy.[58] Since we only have reliable and empirical probabilities for events that have had a long recorded history,[59] some assessors believe that the use of historical risk data for new technologies must result in an underestimation of risk; this is because certain events may not have occurred between the inception of a technology and the end of the period for which the risk information is compiled.[60] Moreover, most claims about the allegedly low probability of catastrophic accidents, as in the case of LNG or nuclear technologies, are predicated on relatively low accident records for these new technologies. There is a problem, however, with making a claim about accident probability on the basis of observed accident frequency. On the one hand, very low values of an accident probability per LNG trip, or per reactor-year, for example, are consistent with an assumed record of zero accidents in 800 voyages, or zero core melts in 17,000 reactor years. On the other hand, a probability as high as 1 in 100 or 1 in 200 would still be consistent with the LNG accident record, just as a probability as high as 1 in 2,000 would be consistent with the nuclear accident record. In other words, even though an accident record may be consistent with very low probability values, this alone "does not prove that the values are low".[61]

Second, the Starr and Whipple view is unable to account for the reasonable controversy, among both Nobel Prize winners and among the American Physical Society (APS), the Environmental Protection Agency (EPA), the Nuclear Regulatory Commission (NRC), and the American Nuclear Society (ANS), over accident probabilities.[62] Reputable assessors affirm that 'good' data on nuclear risk probabilities differs by a factor of 1,000.[63] Moreover, they say, there are a number of difficulties which make nuclear probabilities especially resistant to accurate estimation, e.g., (1) the fact that nuclear accidents involve compound events whose probability depends on estimating sequential component failures; (2) the fact that there is no way to estimate the risk of sabotage or human error;[64] and (3) the fact that some of the most dangerous nuclear risks, e.g., weapons proliferation, are the least amenable to quantification.[65] Rasmussen himself computed the probability of having a Three-Mile-Island-type accident as anywhere from 1 to 250 to 1 in 25,000 reactor years.[66] If this is true, then it suggests that certain accidents are not really 'impossible', because many low probabilities are not believable. For example, the probability for a royal flush is 1 in 646,000 yet, in a card game, the probability is actually much higher, since the probability of cheating is likely to be as high as 1 in 10,000. Likewise, although the probability of a given accident may be only very slight, the higher probability of sabotage or terrorism is likely to increase this number by several orders of magnitude. This means that risks are likely to include so-called "outrageous events" or "rogue events," which are difficult to handle in accident risk assessment.[67] The magnitude of the possibility of such 'events' is quite high since, for example, "at least 80% of marine accidents are caused by human error", which also causes a majority of most transportation and automobile accidents.[68]

Specifically, Starr and Whipple are assuming that given experts are correct in their assessments of nuclear risk. More generally, they are assuming that, in one of the most controversial, untested, and potentially catastrophic areas of technology, it is possible to judge clearly when a risk probability is wholly accurate and when it is not. Starr and Whipple have an appeal to authority, an appeal which (given the history of science) simply does not hold up. They also have used one of the most difficult and multifaceted cases around, nuclear power, in an attempt to clarify and substantiate the difficult point that incorrect judgments of accident probabilities, rather than other errors, play the dominant rule in controversy over technology. Because there are so many parameters, in the nuclear power case, about which informed persons may disagree (e.g., the Price–Anderson Act and insurance compensation; the issue of equity, especially as regards nuclear waste and future generations; and the problem of nuclear economics, especially relative to other means of generating electricity) it is extremely difficult to see how this case illustrates their point about the centrality of disagreement over accident probabilities.

Their point about the importance of probability estimates is especially vulnerable when one recalls that "the nine characteristics hypothesized by various authors to influence judgments of perceived and acceptable risk . . . [are] highly inter-correlated."[69] Involuntary hazards, for example, "tend also to be inequitable and catastrophic".[70] This means that it is especially difficult to determine whether or not society's expressed concern about involuntary risks, for example, is merely an artifact of the excellent correlation between involuntariness and other undesirable risk characteristics. By using the method of observed, rather than expressed, preferences Starr and Whipple seem to me to be unable to affirm, with assurance, that the characteristic of having a low, misperceived probability (rather than some other characteristic highly correlated with this one) is the major cause of controversy over technology. Another way of making the same point is to note that there are numerous allegedly causal explanations, all consistent with the same 'observed' phenomena. Kasper made an analogous observation in his discussion of scientific explanation:

> even the best of epidemiological studies is confounded by the myriad explanations for low level neurobehavioral effects; the same effects attributed to lead may be caused by exposure to low levels of many other trace metals and, indeed, by exposure to the pace and stress of urban life itself. The result is that careful studies yield not proof but only suggestions.[71]

5. THE ARGUMENT THAT THERE IS A DISTINCTION BETWEEN PERCEIVED AND ACTUAL PROBABILITIES

Precisely because they ignore the fact that the characteristic of having a low, misperceived probability is highly correlated with other characteristics, Starr and Whipple are not warranted in asserting that misperceived probabilities *cause* high aversion to nuclear risks. As a consequence, they are not warranted in asserting that persons incorrectly view the risk-cost relationship in a nonlinear way *because* they misperceive the risk probabilities. Rather, as the preceding arguments urged, it is likely that certain risks are viewed nonlinearly because of the low *value* attributed to the benefits of taking the risks.

Moreover, as was also argued, it is epistemologically impossible to provide a clean-cut case for the thesis that catastrophic events, especially those involving relatively new technologies, are *erroneously* perceived as having a high actual probability of fatality.

Hence Starr, Whipple, and others are not warranted in asserting either that certain risks are clearly known to be of a low probability, or that there is always a clear distinction between what they define as *objective* (or 'actual') *societal* risks and *subjective* (or merely 'perceived') *individual* risks, as they say there is.[72] They would have us believe that "misperceived", erroneous probabilities account for the high value placed on safety from certain catastrophic accidents, e.g., at LNG facilities. Using their notion of actual and perceived probabilities to explain apparent anomalies, given this interpretation of 'the linearity assumption', however, presupposes that their version of the actual/perceived distinction will hold up in every case. For many reasons, I think it will not.

First, there are numerous problems of actual risk estimation (prior to any alleged evaluation) which simply do not admit of analytical resolution by experts. As a consequence, they can only be handled by means of the intuitive estimates of individuals. Authors of a recent study done at the Stanford Research Institute recently admitted, for example, that analytical techniques couldn't handle probability estimates for terrorist attacks on nuclear installations. They concluded: "we must rely on expert judgment, quantified using subjective probabilities."[73] One key reason why probabilities are unalterably subjective is that it is impossible to be sure that a given methodology has accurately accounted for all significant possibilities.[74] This is why one official, with the U.S. Environmental Protection Agency, claims that risk "is not like the length of a stick; there is no true 'something' that you are trying to estimate."[75] Likewise, as Hollister pointed out,[76] "Our loss of the three astronauts on the ground at Cape Kennedy demonstrated that even the best no-cost-barred, no-cost-spared, systems analytic approaches in the world cannot anticipate every possibility." This being so, assessors must make value judgments about the factors determining a given probability.

As a consequence, it is often difficult to distinguish actual (objective) and perceived (subjective) risk probabilities. In fact, one of the most famous risk probabilities, widely touted as 'actual', is highly value-laden. This is the reactor-year probability of a core melt in a nuclear plant, 1 in 17,000. As defended and explained in WASH-1400, this probability is notoriously laden with value judgments about the effectiveness of evacuation in the face of a catastrophe,[77] the probability of weather stability,[78] and the Gaussian Plume rise of radioactivity.[79] The problem is *not* that the WASH-1400 probabilities, or other actual probabilities (as given by experts) are value-laden, but that they are apparently not recognized as such by Starr, Whipple, and others, who attempt to draw a sharp line between actual (objective) and perceived (subjective) probabilities.

Second, it is well-known that individuals discount risks in both the space and time dimension. They are more concerned with events in their own neighborhood than with those in either a distant time or a remote place.[80] In fact, if societal (or 'actual') risk is taken to be the risk to that refined subset of persons closest to a problem, and if the population is divided into finer and finer risk subsets (e.g., those living within 50 miles of an LNG facility; those bearing higher risks from industrial radiation emissions because of previous medical exposures), then the alleged difference between their definitions of *actual* societal risk and *perceived* individual risk disappears.

Third, their version of the actual/perceived risk distinction is dependent upon the assumption that only the *average* of the aggregated risk probabilities represents the correct, or 'actual', probability. Meanwhile, all accurate risk probabilities for particular subsets of persons are assumed to be merely 'perceived', non-analytical, subjective, individual, and potentially incorrect.

Consider the consequences, to which the acceptability of such average probabilities can lead. Starr, for example, argues repeatedly in his work that "the highest level of acceptable risks which may be regarded as a reference level is determined by the normal U.S. death rate from disease."[81] Based on this criterion, he suggests that we ought not necessarily to have been overly concerned about the numerous deaths occasioned by the war in Vietnam. Writing during the peak of the conflict, he coolly affirmed that "the related risk, as seen by society as a whole, is not substantially different from the average nonmilitary risk from disease."[82] Only in one limited sense is his point correct. Part of what is questionable about it is whether the normal death rate from disease provides a sufficient criterion for judging a risk's acceptability, and whether that criterion ought to be defined merely in terms of the actual probability of fatalities. Also questionable is the interpretation which Starr places on risk acceptability as a direct result of the group whose risk probabilities he chose to aggregate, then average. If he had compared the military risk to U.S. males in the 18 to 24 age group, instead of to males and females of all ages, then obviously he could not have concluded that the Vietnam risk was "not substantially different from the average nonmilitary risk from disease."[83] Likewise, because they define 'actual' (or correctly perceived) risk, by aggregating and then averaging all risk probabilities, assessors' distinguishing between their senses of actual, and merely perceived, risks is question-begging. Once the relevant subsets of persons are considered, their allegedly perceived risk may be more correct, more 'actual', than that defined by the assessors as actual.

The same methodological failing, regarding a risk as negligible merely because it is measured against the *average* risk to all persons from all activities, occurs throughout much risk assessment literature. Typically, effects of low-level radiation are said to be negligible. They are so small that they are masked by other environmental factors and are perceived as insignificant. Interestingly, however, the same judgment of insignificance could be made about other causes of death (such as murder by handguns) "which we do regard as of some significance. If we had no way of distinguishing death by murder from death by natural causes, [then] the death rate from murder could increase manyfold before it became noticeable as an increase in the mortality from all causes."[84]

The point of my argument here is twofold. First, risk probabilities for various subclasses of persons are essential to accurate and equitable risk assessment. Second, once one begins to consider the actual risk faced by particular classes of individuals, then one has already begun to collapse the risk assessors' distinction between perceived or individual, and actual or societal, risk probabilities. Although there is often a valid distinction between subjective and objective probabilities, Starr, Whipple, and others have *defined* 'actual' (objective) versus 'perceived' (subjective) probabilities is a highly doubtful and question-begging way, viz., in terms of averages, and have then relied on this definition to defend their linearity assumption. As a consequence, it seems misleading for Starr, Whipple, and others to attempt to explain the anomalies with regard to this interpretation of 'the linearity assumption' merely by saying that perceived, not actual, risk probabilities, as they define them, cause the public the view risks in nonlinear and "inconsistent" ways. Their notion of *'actual'* probabilities is nothing more than probabilities calculated on the basis of their own theoretical assumptions.

Starr, Whipple, and other risk assessors thus attempt to gain support for the linearity assumption by using a very restricted notion of 'actual' (objective) probability. In assuming that theoretical and statistical probabilities are indeed actual or real, they have defined 'actual' probabilities in a way that makes the linearity assumption true by fiat. A more accurate way to think of probabilities might be as *actual* or *real, theoretical, statistical*, and *perceived.* One might define them as follows. *Real* probabilities are those which, after the occurrence of all the events in question, are known with certainty to be the case. *Theoretical* probabilities are those which before the occurrence of all the events in question are estimated to be the case, on the bases of certain mathematical, factual, epistemological, and scientific assumptions. *Statistical* probabilities are those which, before the occurrence of all the events in question, are thought to be the case, solely on the basis of past frequencies. *Perceived* probabilities are those which, before or after the occurrence of all the events in question, are thought by the public to be the case, for whatever reason. From these definitions, it is clear that only *real* probabilities are 'objective', in the sense of being accurate or being 'verified' by actual occurrences. Because later experience could prove that certain theoretical, statistical, or perceived probabilities were erroneous, it is likewise clear that they could be 'subjective', in the sense of not being consistent with actual occurrences. Since even proponents of various probability theories admit that their results are subjective,[85] it is puzzling that Starr, Whipple, and others view theoretical probabilities (based on analysts' calculations) as 'actual' or objective. As I have argued, these probabilities are sometimes as subjective as perceived probabilities. If so, then Starr, Whipple, and others cannot assume, in a given case, that these theoretical probabilities are actual and that they support the linearity assumption.

However, even if Starr, Whipple, and others were correct in claiming that low-probability events are typically misperceived as likely (Sections 3 and 4), and that there is a clear distinction between *actual* (including theoretical and statistical) as opposed to merely perceived, probabilities (Section 5), these facts alone would not prove the observed societal aversion to uncompensated public risks is caused merely by misperceived probabilities of fatalities. As was suggested earlier, many other factors, such as the distribution of the risk, or its trade-off with benefits, could equally well explain the anomalies. Moreover, even if there were no counterexamples to the fatality interpretation of 'the linearity assumption', and even if the public unanimously judged that the value of risk avoidance was always proportional to the actual probability of fatality, neither this claim, nor any other argument based on consensus, would establish the *correctness* of the fatality interpretation, although it might establish the *political desirability* of accepting it.

6. A RECONSIDERATION

If this is true, then assessors ought to reexamine the claim against them, viz., that their tenet (3) [that there is a *linear relation* between societal risk (expressed in terms of probabilities) and the value of risk avoidance] is incorrect, especially in the cases in which one is evaluating low-probability catastrophic risks. At this point, however, one might wonder how Starr, Whipple, *et al.* could hold an apparently simplistic and false thesis (the linearity assumption). Do they really subscribe to this view?

There are three reasons why Starr, *et al.*, do appear to subscribe to the assumption in question. For one thing, they state this claim and they illustrate it on their graphs of risk evaluation. Starr's and Whipple's first figure, in their 1980 *Science* article, shows a graph of the societal value of risk avoidance; the value is linear over the range (0–1) of probability of fatality. They state explicitly, "as Fig. 1 illustrates . . . the presumption of a linear relation between risk and the cost of that risk may be quite valid."[86]

A second reason why they and other assessors appear to subscribe to the linearity assumption is that it is a clear presupposition of their criticism of those who have more aversion to certain low-probability risks than to higher-probability ones. If they did not make this assumption about linearity, then they would not be able to conclude that the counterexamples were the result of misperceived probabilities of fatality, rather than a product of considering some other factor (e.g., another unit-cost burden or a particular benefit) relevant to the risk outcomes. Therefore they must subscribe to the linearity assumption.

Third, Whipple himself has admitted, in a private conversation, that he and other risk assessors do accept this assumption. He maintains, however, that it is only a minor part of their theory.[87] (This latter claim is surely false, for reasons just mentioned. Without the linearity assumption, risk assessors would not be able to compare probabilities of fatality, *across different risks*, and claim that the cost of each risk was *proportional* to its probability of fatality.[88] In other words, Starr and others are able to talk about proportionality and to obtain commensurability among risks, a necessary aspect of their systematic approach, only by making the linearity assumption. For this reason, the assumption is central to their enterprise. Of course, if they did not claim that the cost of a risk was proportional to its probability of fatality, then the linearity assumption would not be necessary to their obtaining commensurability among risks.)

Admittedly, however, understanding the assumptions of risk assessors is no easy task. I am troubled, for example, by the fact that, in recent articles, Starr, Rudman and Whipple[89] illustrate that, for societal risks, the value of risk avoidance is linear over a risk range (0–1) of probability of fatality, but that, in the same articles, they claim that "a complete risk–benefit decision requires that the relative social cost of the risk be compared with the associated benefit" and that "the risk–benefit analysis weighs the social cost against the benefits to gain insights into the criteria for acceptability."[90] If these statements are true, then how can probability of fatality and value of risk avoidance be linearly related, as they also claim? What happens to *benefits* in such a relationship? Clearly, the introduction of a second independent variable (benefits) makes linearity impossible. The assertion that one variable is a linear function of another seems to entail that the value of the first variable is independent of any other variable but the second. (If Y is a linear function of X, then it is independent of Z, with the exception of any Z for which X is a function of Z (since transitivity would require that Y be a function of Z).)

If this is so, then what, precisely, are Starr, Whipple, *et al.* asserting about the relationship between risks, benefits, and the value of risk avoidance? If we are not to assume that they are inconsistent, and I think that they are not, then their remarks (previously quoted) indicate that they wish to say that, although consideration of benefits is essential "to gain insights into the criteria for acceptability", benefits *per se* are excluded from their measure of the "value of risk avoidance". Moreover, they appear to believe that although benefits are excluded from the measure of the "value of risk avoidance", consideration of them is important for "a complete risk–benefit decision". This suggests that knowing the "value of risk avoidance" does not provide sufficient grounds for making "a complete risk–benefit decision". In turn, this suggests that what Starr, Whipple, *et al. say* about the importance of benefits is not necessarily borne out by what they *do* to include them in their methodology. This interpretation of their position is plausible, I think, especially because, in considering particular cases, they clearly affirm: "an assessment of benefits is not necessary for the general analyses of energy technologies."[91] In other words, although they pay lip service to the importance of benefits in evaluating risks, since their method provides no way of accommodating them, benefits are often in practice ignored. A more sympathetic interpretation is that Starr, Rudman, Whipple, *et al.* would admit that their analysis is simplistic and still at a very imcomplete stage. In fact, Starr himself recognized that many of his proposals were simplistic. In a personal communication to Otway, he said: "I consider the hypothesis which I have suggested as both very preliminary and simplistic in what must eventually be a more complicated analytic process."[92] Likewise, in his early classic article in *Science*, Starr admitted that his definition of risk in terms merely of "statistical probability of fatalities" was "for the sake of simplicity in this initial study".[93] Nevertheless, in articles published eleven years later, Starr and others clearly still assume that risk is defined merely in terms of probabilities, since they criticize as irrational those who view some higher probability risks as more acceptable than lower probability ones.[94] Hence, assessors' continuing adherence to the linearity assumption suggests that risk assessment needs to be purged of a number of questionable tendencies, all associated with this assumption.

7. SOME INSIGHTS ABOUT RISK ASSESSMENT

What are these questionable tendencies? At least three come to mind. *First*, because of their reliance on the assumption of a linear relationship between probability of fatality and the value of risk avoidance, and because of their frequent failure to consider parameters other than probability of fatality in estimating risk, many (if not most) assessors underestimate the value component in risk determination. They assume that any preference for a risk, whose probability-of-fatality is statistically higher than that of an alternative, is a result of misperceived probabilities, not a result of a given value system. This failure to recognize the value components of allegedly objective probability estimates goes hand in hand with the assumption that controversy over technology, arises in large part because the public does not accept experts' allegedly objective risk estimates. Recognizing the value component in these estimates is the first step toward avoiding the simplistic dichotomy between actual/expert and perceived/layman's probability estimates. Obviously both estimates are perceived evaluatively, even though the expert likely has a more accurate factual knowledge. If this is not recognized, however, then real negotiation over controversial technological policies will be impossible. This is because the first step in negotiation is mutual recognition of the authentic sources of conflict, viz., conflict over values.

One reason for this lack of recognition of the value component in controversy over technology is that people too often attempt to define ethical and political issues as merely technical ones. They assume, incorrectly, that agreement about technical matters is sufficient for resolving normative disputes. Apparently they make this assumption because they are afraid of damaging "the scientific pretenses of their work".[95] As a consequence, the emphasis placed on the importance of abstract science helps both to disguise the often exploitative way in which technology is used, and to condone a passive acceptance of the status quo. It further allows assessors to dismiss as irrational or unscientific (as Okrent, Starr, Whipple, Maxey, Cohen, and Lee have done) any attempts to challenge our contemporary ethical or political values.[96] In other words, "the use of supposedly objective models of human and social behavior serves to legitimate the imposition of social policy".[97] This is because allegedly subjective considerations, especially about values, could easily be dismissed as nonanalytic and intuitive, and based only on perceived, as opposed to actual, risk.

A *second* questionable tendency associated with the fatality interpretation of the linearity assumption is assessors' reliance on allegedly *objective* probabilities. In explaining counterexamples to their models as those in which probabilities are misperceived, assessors appear to believe that probabilities can be determined in a wholly objective manner in every situation. Häfele, Okrent, Whipple, and Lowrance, for example, all postulate that the probability component, $(p_i(x_i))$ of the utility function, $U = \Sigma\, p_i(x_i) \cdot u_i(x_i)$ is "objective", even though the u_i is subjective or evaluational.[98] Starr and Whipple likewise assume that there is an objective, wholly accurate probability which is capable of being calculated, before the occurrence of all the events in question. They propose that controversy over technology is largely a matter of conflict between those who know *actual* risk probabilities and those who know merely *perceived* risk probabilities.[99] Apart from whether their point about the cause of controversy is true, their theory errs in ignoring the fact that all risk probabilities, even allegedly 'actual' ones, involve value judgments, and that "experts' risk assessments are also susceptible to bias, particularly under-estimation due to omitting important pathways to disaster."[100] This is because there are a number of different uncertainties which jeopardize the alleged objectivity of given probability estimates or calculations. Fairley summarizes the errors arising in probability estimates as those of reporting, extrapolation, speculation, definition, and theory.[101] Speculative estimates, for example, might err because of "inclusion uncertainty", i.e., uncertainty about relevant consequences or about identification of all recipients of exposure coverage, or because of "specification uncertainty", i.e., uncertainty about whether given probability pathways are mutually exclusive or exhaustive. At a more basic level, there are also parameters (e.g., the probability of sabotage) which are in principle uncertain.[102] In addition, there are numerous *modelling* errors and assumptions, often resulting from oversimplification of pathways and consequences, as well as *measurement* errors and assumptions, often in obtaining exposure magnitudes.[103]

Perhaps in part as a consequence of these various errors in estimating probabilities, numerous scientists and risk assessors chronically misjudge sample implications and place too much weight on samples which are too small. They also often fall victim to a number of judgmental biases, such as the "availability bias" (judging a probability on the basis of one's ability to imagine or remember relevant instances) and the "anchoring bias" (judging a probability on the basis of how well it confirms one's original judgments (anchor), apart from what new information may reveal).[104]

A *third* questionable tendency associated with the linearity assumption is that risk assessors appear to be overly concerned with what they regard as erroneous public decisions. The root of this concern is that they tend to view allegedly objective characterizations of risk, illuminated by experts' calculations, as somehow more real or more valid than the risk evaluations of the public. If my remarks have been correct, however, then the evaluations of the public are often quite reasonable, particularly if they are based on important evaluational assumptions often ignored by assessors. Because experts often define risk merely in terms of probability of fatality and consequently neglect these concerns, they fail to attack one of the real problems of risk assessment: how to make the decisionmaking process more democratic. This, however, is surely questionable, at least in our society. We decided long ago in the policy process to stake our claim with democratic procedures informed by expert opinion, rather than solely with expert determination. We believe, for example, that "the process by which a decision is reached is more important than its factual correctness",[105] and in some cases this may hold true for science policy as well as for politics. We also believe "that democratic acquittal of a criminal defendant known to be guilty is better than conviction solely on the basis of a scientifically objective, but undemocratically determined guilt"; as Green puts it: we "have chosen to tolerate error and wrong decisions rather than to establish a benevolent dictatorship that could ascertain and implant the truth reliably and authoritatively."[106]

Reliance solely on expert predictions, as opposed to public evaluations informed by expert opinion, is also likely to generate more pro-technology bias because decisionmakers "will be more reluctant to prevent risks that are uncertain than to approve projects whose benefits are speculative."[107] Reliance on expert assessments alone also tends to be associated with the belief that political processes are irrational.[108] This means that policymakers might use techniques of propaganda and indoctrination, rather than public discussion, analysis, and information sharing, to persuade the public that they are wrong, rather than to try to understand the values to which the public often subscribes. This, in turn, can only lead to an erosion of trust and respect between the two groups (the people versus the experts) and to an increased polarization of the 'two cultures'.[109]

Contrary to the assumption that the concerns of allegedly intuitive assessors could well be ignored, Thomas Jefferson warned that the only safe locus of societal power was in the people. He wrote: "I know of no safe depositor of the ultimate powers of the society but the people themselves; and if we think them not enlightened enough to exercise their control with a wholesome discretion, the remedy is not to take it from them, but to inform their discretion."[110]

NOTES

1 Many of these same observations are made by P. F. Ricci and L. S. Molton, 'Risk and Benefit in Environmental Law', *Science* **214** (4525), (4 December 1981), 1096–1100.

2 Starr and Whipple, for example, claim that "historically revealed social preferences and costs are sufficiently enduring to permit their use for predictive purposes", and that "in such historical situations a socially acceptable and essentially optimum tradeoff of such values has been achieved" (C. Starr, 'Social Benefit versus Technological Risk', *Science* **165** (3899), (19 September 1969), 1232–1233; hereafter cited as: Starr, Social. See also C. Starr, 'General Philosophy of Risk–Benefit Analysis', in H. Ashley, R. Rudman, and C. Whipple (eds.), *Energy and the Environment: A Risk–Benefit Approach*, Pergamon, New York, 1976, p. 3; hereafter cited as: Starr, RBA, in *Energy*.) As Starr himself recognizes, these assumptions may be doubtful, since his method doesn't distinguish what is "best" from what is "traditionally acceptable" (Starr, Social, p. 1232; see Starr, *Energy*, p. 3). What he fails to recognize, however, is that the method cannot distinguish even what is "traditionally acceptable". Risks are obviously not acceptable just because society has encountered or taken them. Some may have endured through

ignorance, or because of political impotence, or the unavailability of alternatives. Moreover, some risks have only been discovered years after exposure to them because of the difficulties associated with data gathering.

[3] C. Starr, *Current Issues in Energy*, Pergamon, New York, 1979, p. 23; hereafter cited as: *CIE*. C. Starr and C. Whipple, 'Risks of Risk Decisions', *Science* **208** (4448), (6 June 1980), 1116; hereafter cited as: Risks. C. Starr, R. Rudman, and C. Whipple, 'Philosophical Basis for Risk Analysis', *Annual Review of Energy* **1** (1976), 640–641; hereafter cited as: Philosophical. See also L. Philipson, 'Panel on Accident Risk Assessment'. In Mitre Corporation, *Symposium/Workshop . . . Risk Assessment and Governmental Decision Making*, the Mitre Corporation, McLean, Virginia, 1979, p. 385; hereafter cited as: Philipson, Panel, and Mitre, Symposium.

[4] B. Cohen and I. Lee, 'A Catalog of Risks', *Health Physics* **36** (6), (1979), 707; hereafter cited as: Catalog.

[5] Starr and Whipple, Risks, pp. 1115–1116; Starr, Rudman, Whipple, Philosophical, pp. 636–637.

[6] C. Comar, 'Risk: A Pragmatic De Minimis Approach', *Science* **203** (4378), (1979), 319; hereafter cited as: Risk. See also Starr, *CIE*, p. 12.

[7] D. Okrent, 'Comment on Societal Risk', *Science* **208** (4442), (1980), 372; hereafter cited as: Risk.

[8] S. Gibson, 'The Use of Quantitative Risk Criteria in Hazard Analysis', in D. Okrent (ed.), *Risk–Benefit Methodology and Application*, UCLA School of Engineering and Applied Science, Los Angeles, 1975, p. 599; hereafter cited as: Gibson, Use, in Okrent, *RBM*.

[9] U.S. Nuclear Regulatory Commission, *Reactor Safety Study: An Assessment of Accident Risks in U.S. Commercial Nuclear Power Plants*, Report No. WASH-1400, U.S. Government Printing Office, Washington, D.C., 1975, p. 37; hereafter cited as: NRC, WASH-1400.

[10] See D. Braybrooke and P. Schotch, 'Cost–Benefit Analysis under the Constraint of Meeting Needs'. Working Paper.

[11] Cohen and Lee, Catalog, p. 720. See also J. Gardenier, 'Panel: Accident Risk Assessment', in Mitre, *Symposium*, pp. 399, 467.

[12] L. Lave, 'Discussion', in Mitre, *Symposium*, p. 541. See Starr and Whipple, Risks, p. 1116; D. Okrent, 'Panel: Use of Risk Assessment', in Mitre, *Symposium*, p. 663; D. Bazelon, 'Risk and Responsibility', *Science* **205** (4403), (1979), 278; hereafter cited as: Bazelon, Risk.

[13] See, for example, Cohen and Lee, Catalog, p. 707; W. Häfele, 'Energy', in C. Starr and P. Ritterbush, *Science, Technology, and the Human Prospect*, Pergamon, New York, 1979, p. 139; hereafter cited as: Häfele, Energy, in Starr and Ritterbush, *Science*. See also M. Maxey, 'Managing Low-Level Radioactive Wastes', in J. Watson (ed.), *Low-Level Radioactive Waste Management*, Health Physics Society, Williamsburg, Virginia, 1979, pp. 410, 417.

See also C. Starr, 'Benefit–Cost Studies in Sociotechnical Systems', in Committee on Public Engineering Policy, *Perspectives on Benefit–Risk Decision Making*, National Academy of Engineering, Washington, D.C., 1972, pp. 26–27; hereafter cited as Starr, B–C, in Committee, *Perspectives*. Finally, see Lave, 'Discussion', in Mitre, *Symposium*, p. 484.

[14] Although society's intuitive evaluations of given risks do not provide sufficient grounds for arguing that these risks *ought* to be evaluated in a certain way, many risk assessors maintain that *correct* societal evaluations are consistent with tenet (3) and the linearity assumption. Hence, in response to alleged counterexamples to this assumption, assessors maintain that high public aversion to certain low probability risks does *not* provide a counterexample to the thesis that "actual" risk probabilities and the value of risk avoidance are linearly related, since the public's aversion in such cases is generated by "perceived" (i.e., incorrect) risk probabilities and not "actual" ones.

[15] See W. Rowe, *An Anatomy of Risk*, Wiley, New York, 1977, p. 962; hereafter cited as: *Risk*.

[16] B. Fischhoff, P. Slovic, and S. Lichtenstein, 'Facts and Fears', in R. Schwing and W. Albers (eds.), *Societal Risk Assessment*, Plenum, New York, 1980, p. 207; hereafter cited as: Fischoff, *Risk*, in Schwing and Albers, *Risk*.

[17] A. J. Van Horn and R. Wilson, 'The Status of Risk–Benefit Analysis', discussion paper, Harvard University, Cambridge, Massachusetts, Energy and Environmental Policy Center, 1976, p. 18; hereafter cited as: SRBA.

[18] P. D. Pahner, 'The Psychological Displacement of Anxiety: An Application to Nuclear Energy', in D. Okrent, *Risk–Benefit Methodology and Application*, UCLA, School of Engineering and Applied Science, 1975, p. 575; hereafter cited as: Pahner, Psychological, in Okrent, Methodology.

[19] Van Horn and Wilson, SRBA, p. 19.

[20] Starr and Whipple, Risks, pp. 1115–1116.

[21] R. Zeckhauser, 'Procedures for Valuing Lives', *Public Policy* **23** (4), (1975), 442; hereafter cited as: Procedures.

[22] See S. Zivi, 'Panel: Public Perceptions of Risk', in Mitre, *Symposium*, p. 574, who argues that societal stress is "not proportional to the number of lives lost".

[23] Starr, B–C, pp. 26–27.

[24] Starr, B–C, pp. 26–27.

[25] Starr and Whipple, Risks, pp. 1116–1117; Starr, *CIE*, pp. 16–17; H. Hollister, 'The DOE's Approach to Risk Assessment', in Mitre, *Symposium*, p. 57. See R. Kasper, 'Panel: Use of Risk Assessment', in Mitre, *Symposium*, p. 610, for a critique of this view.

[26] Starr and Whipple, Risks, p. 1117.

[27] Starr and Whipple, Risks, pp. 1116–1117; C. Starr, B–C, pp. 26–27; Starr, *CIE*, pp. 16–17.

28 NRC, WASH-1400, pp. 187–221, esp. fig. A3.2. See also Rowe, *Risk*, p. 320; J. Yellin, 'Judicial Review and Nuclear Power', *George Washington Law Review* **45** (5), 1977), 978; hereafter cited as Yellin, Judicial. See also W. Häfele, 'Benefit–Risk Tradeoffs in Nuclear Power Generation', in H. Ashley, R. Rudman, and C. Whipple (eds.), *Energy and the Environment*, Pergamon, New York, 1976, p. 162; hereafter cited as: Häfele, BRT, in Ashley, Rudman, and Whipple, *EE*.
29 Starr and Whipple, Risks, p. 1116.
30 Starr, B–C, pp. 26–27.
31 Starr, B–C, pp. 26–27.
32 Starr and Whipple, Risks, p. 1116; Starr, *CIE*, pp. 16–17.
33 H. Otway, 'Risk Assessment and the Social Response to Nuclear Power', *Journal of the British Nuclear Engineering Society* **16** (4), (1977), 331; hereafter cited as: Otway, Risk.
34 Otway, Risk, p. 331.
35 Otway, Risk, p. 332.
36 Otway, Risk, p. 332.
37 E. Lawless, *Technology and Social Shock*, Rutgers University Press, New Brunswick, N.J., 1977, p. 512; hereafter cited as: *Technology*.
38 Lawless, *Technology*, pp. 497–498.
39 Lawless, *Technology*, pp. 349–356, 434–435.
40 Lawless, *Technology*, pp. 434–435.
41 Lawless, *Technology*, p. 490.
42 B. Fischhoff, P. Slovic, S. Lichtenstein, S. Read, and B. Combs, 'How Safe Is Safe Enough?' *Policy Sciences* **9** (2), (1978), 150; hereafter cited as: Fischhoff, Safe.
43 In arguing that the nuclear debate is primarily over nuclear accident probabilities, Starr and Whipple appear to dismiss the importance of the question of whether the nuclear benefit trade-off is worth the risk. In other parts of their work, however, they clearly state that "risks and benefits are not evaluated independently" (Risk, p. 1117). I agree, but if risks and benefits are not evaluated independently, how can they be so sure that the nuclear debate is primarily over the risk probabilities, rather than over whether the benefit is worth the risk?
44 Fischhoff, *Risk*, p. 192.
45 Fischhoff, *Risk*, p. 192.
46 Fischhoff, Safe, pp. 140–142; *Risk*, p. 202. See also R. Kasper, 'Perceptions of Risk', in Schwing and Albers, *Risk*, p. 75; hereafter cited as: Perceptions.
47 Fischhoff, Safe, pp. 148–149; see also H. Green, 'Cost–Benefit Assessment and the Law', *George Washington Law Review* **45** (5), (1977), 909–910; hereafter cited as: CBA.

[48] Fischhoff, *Risk*, p. 208; Rowe, *Risk*, p. 290.

[49] Yellin, Judicial, p. 992.

[50] Yellin, Judicial, p. 987.

[51] Yellin, Judicial, pp. 983–984.

[52] Yellin, Judicial, pp. 987–988.

[53] J. G. Palfrey, 'Energy and the Environment', *California Law Review* **74** (8), (December 1974), 1377; hereafter cited as: Palfrey, EE.

[54] James Lieberman, 'Generic Hearings', *Atomic Energy Law Journal* **16** (2), (Summer 1974), 142; hereafter cited as: Generic.

[55] Starr and Whipple, Risks, p. 1116.

[56] U.S. NRC, *Reactor*, pp. 40, 96, 97, 244; see Lieberman, Generic, 1976, p. 270.

[57] U.S. NRC, *Reactor*, p. 15.

[58] D. Okrent and C. Whipple, *Approach to Societal Risk Acceptance Criteria and Risk Management*. Robert No. PB-271264, U.S. Department of Commerce, Washington, D.C., 1977, p. 10; hereafter cited as: *Approach*.

[59] W. Lowrance, 'The Nature of Risk', in Schwing and Albers, *Risk*, p. 6; hereafter cited as Nature.

[60] Rowe, *Risk*, p. 264.

[61] W. Fairley, 'Criteria for Evaluating the "Small" Probability', in Okrent, *RBM*, p. 425; hereafter cited as: Fairley, Criteria.

[62] U.S. NRC, *Reactor*, Appendix XI, pp. 2–1 through 2–14; see also W. Häfele, 'Benefit Risk Tradeoffs in Nuclear Power Generation', in H. Ashley, R. Rudman, and C. Whipple (eds.), *Energy*, pp. 159–160; hereafter cited as: BRT. See also Lieberman, Generic, pp. 250–255.

[63] Yellin, Judicial, p. 988.

[64] Zeckhauser, Procedures, p. 445; Committee, *Perspectives*, p. 10.

[65] Lowrance, Nature, p. 11.

[66] N. C. Rasmussen, 'Methods of Hazard Analysis and Nuclear Safety Engineering', in T. Moss and D. Sills (eds.), *The Three Mile Island Nuclear Accident*, New York Academy of Sciences, New York, 1981, p. 29.

[67] Fairley, Criteria, pp. 406–407.

[68] Philipson, Panel, p. 246.

[69] Fischhoff, Safe, pp. 144, 149.

[70] Fischhoff, *Risk*, pp. 144, 149.

[71] R. Kasper, 'Perceptions of Risk and Their Effects on Decision Making', in Schwing and Albers, *Risk*, p. 73; hereafter cited as: Perceptions.

[72] Starr and Whipple, Risks, pp. 1115–1117. See also J. Burnham, 'Panel: Use of Risk Assessment', in Mitre, *Symposium*, p. 678.

[73] A. Lovins, 'Cost–Risk–Benefit Assessment in Energy Policy', *George Washington Law Review* **45** (5), (1977), 926; hereafter cited as: Lovins, CRBA.

[74] Philipson, Panel, pp. 415, 419.

[75] T. Feagan, 'Panel: Human Health Risk Assessment', in Mitre, *Symposium*, p. 291.

[76] H. Hollister, 'The DOE's Approach to Risk Assessment', in Mitre, *Symposium*, p. 50.

[77] NRC, WASH-1400, pp. 118, 186, 245.

[78] NRC, WASH-1400, p. 239.

[79] NRC, WASH-1400, pp. 108–109, 186, 239, 245–246.

[80] Rowe, *Risk*, p. 130.

[81] Starr, *CIE*, p. 14; Starr, *et al.*, Philosophical, p. 630.

[82] Starr, Social, p. 1235.

[83] Starr, Social, p. 1235.

[84] Lovins, CRBA, pp. 921–922.

[85] L. Savage, *The Foundations of Statistics*, Wiley, New York, 1954, pp. 56–60.

[86] Starr and Whipple, Risks, pp. 1115–1116.

[87] Professor Sheldon Krimsky of Tufts University discussed this matter with C. Whipple and reported his response. I am grateful to Professor Krimsky for his information on this point.

[88] I am grateful to Professor Joseph Agassi of Boston University for pointing out the centrality of the linearity assumption to RCBA.

[89] Starr, Rudman, Whipple, Philosophical, pp. 636–637; Starr and Whipple, Risks, p. 1115.

[90] Starr, Rudman, Whipple, Philosophical, p. 638.

[91] Starr, Rudman, Whipple, Philosophical, p. 638.

[92] Cited by Rowe, *Risk*, p. 303.

[93] Starr, Social, p. 1234.

[94] Starr and Whipple, Risks, pp. 1116–1117.

[95] H. Stretton, *Capitalism, Socialism, and the Environment*, Cambridge University Press, Cambridge, 1976, p. 51; hereafter cited as: *CSE*.

[96] See Dickson, *The Politics of Alternative Technology*, Universe Books, New York, 1975, p. 189; hereafter cited as: *PAT*.

[97] Dickson, *PAT*, and Stretton, *CSE*.

[98] Häfele, RBT, p. 181; Okrent and Whipple, *Approach*, pp. 1–9; W. Lowrance, *Of Acceptable Risk*, Kaufman, Los Altos, 1976, p. xx. See also Rowe, *Risk*, pp. 3, 5, and Kasper, Perceptions, p. 72.

[99] Starr and Whipple, Risks, p. 1117.

[100] Fischhoff, *Risk*, p. 211.

[101] Fairley, Criteria, pp. 435–436.

[102] Lovins, CRBA, pp. 925 ff., and Kasper, Perceptions, p. 74.

[103] Rowe, *Risk*, p. 122.

[104] For related points, see S. Koreisha and R. Stobaugh, 'Appendix: Limits to Models', in *Energy Future*, ed. Stobaugh and Yergin, Random House, New York, 1979, pp. 234–265.

[105] Lovins, CRBA, p. 941.

[106] H. Green, 'Cost–Benefit Assessment and the Law: Introduction and Perspective', *George Washington Law Review* **45** (5), (1977), 910; hereafter cited as: CBA.

[107] Green, CBA, p. 901.

[108] Committee on Public Engineering Policy, *Perspectives on Benefit–Risk Decision Making*, National Academy of Engineering, Washington, D.C., 1972, p. 7; hereafter cited as: Committee, *Perspectives.*

[109] R. Kasper, 'Perceptions of Risk and Their Effects on Decisionmaking', in Schwing and Albers, *Risk*, pp. 77–78; hereafter cited as: Kasper, Perceptions.

[110] Cited in D. Bazelon, 'Risk and Responsibility', *Science* **205** (4403), (1979), 277–280.

CHAPTER SEVEN

WHERE WE GO FROM HERE

1. INTRODUCTION

If the discussion of the preceding chapters is correct, then risk assessment needs to be improved in some significant ways. In general, it needs to avoid simplistic approaches to problems of safety, and to address the complex epistemological, logical, scientific, and ethical problems raised by questions of evaluating societal risks. In particular, in terms of the methods of revealed preferences, risk assessors could take some first steps toward improvement of their methods and techniques by foregoing appeal to the commensurability presupposition (Chapter 3), the theory of the compensating wage differential (Chapter 4), the probability-threshold position (Chapter 5), and the linearity assumption (Chapter 6). Instead, they need to investigate the ethical and methodological constraints which, in a particular situation, determine whether these appeals are philosophically defensible.

Of course, although ethical and methodological analyses could do much to improve the current practice of risk assessment and the scientific theory underlying it, philosophical scrutiny alone is inadequate to accomplish the reform which appears needed in risk assessment and risk management. Particularly with respect to risk management, widespread legal and political changes need to be implemented in order to speak to some of the difficulties raised earlier in this volume. For example, to address the problems associated with the double standard for occupational and worker exposure to risk (Chapter 4), all governments ought to attempt to establish international standards for health and safety, perhaps through the International Labor Organization (ILO).[1] Other desirable risk-management policies, likely to alleviate some of the difficulties mentioned earlier in Chapter 4, include passage of right-to-know legislation for workers in all countries, so that employers everywhere would be forced to reveal the chemical names and hazards of all substances used on the job. Workers also ought to have access to their medical and exposure records and to hygienists of their own choosing, rather than merely to the company medical doctor.[2] However, since the primary concern of this volume is not with political and legal reform of existing risk-management policies, but with philosophical analysis of risk-assessment methods, my two proposals for change have to do chiefly with improving the current methods and techniques of risk analysis and with suggesting new ones. Let's examine each proposal.

2. AMENDING CURRENT METHODS AND TECHNIQUES OF RISK ANALYSIS

Although earlier chapters in this volume do not provide an exhaustive discussion of the numerous methodological problems facing contemporary risk assessors, addressing the four difficulties analyzed earlier is a first step toward what I hope will be an ongoing and exhaustive discussion in the future. As was pointed out in Chapter 3, one of the most controversial methodological assumptions of contemporary risk assessors is what I have called 'the commensurability presupposition'. This is the assumption that the marginal cost of saving lives, across opportunities, ought to be the same. If my arguments made earlier are correct, then assessors ought to reject universal employment of this presupposition. Instead, on a case-by-case basis, they ought to employ the three factual and the two ethical criteria (developed in Sections 3.2–3.3.2 of Chapter 3) in order to determine whether, in a given instance, one ought to assume that the marginal cost of saving lives, across opportunities, ought to be the same.

Another highly controversial presupposition widely employed among contemporary risk assessors emerges straightforwardly out of classical economic theory. This is what I call the assumption of the 'compensating wage differential'. If my arguments in Chapter 4 are correct, then current risk analyses err if they presuppose that the double standard for public and occupational risk is justified by an appeal to this assumption. Hence, unless they can come up with some new justification, risk assessors ought to reject their adherence to a double standard for public and occupational risk and to develop a uniform standard for risk exposure.

Despite the fact that it was formally adopted by various U.S. regulatory agencies in 1983 (see Section 2 of Chapter 5), what I call the 'probability-threshold position' remains controversial. This is the view that average, annual, individual probabilities of fatality smaller than 10^{-6} are negligible and therefore ought to be ignored by risk assessors. If my arguments in Chapter 5 are correct, however, then the probability-threshold position ought to be abandoned, both because it rests on the erroneous assumption that probability alone is a sufficient condition for evaluating risks and because it is postulated on the assumption that magnitude of probability (apart from ethical considerations such as compensation, equity of risk distribution, and risk–benefit trade-off) is the sole basis for determining the acceptability of a given risk. Because of these flaws in the probability-threshold position, I argued that assessors ought not to employ it in analyzing risks. In the next section of this chapter, I will briefly describe a methodological alternative to the probability-threshold position. This alternative consists of using weighted risk–cost–benefit analysis (RCBA) to evaluate various technological and environmental risks.

Like those who subscribe to the probability-threshold position, advocates of the linearity assumption, which I discussed in

Chapter 6, err in assuming that magnitude of probability is a sufficient basis for determining the acceptability of a given risk. (The linearity assumption is the view that there is a linear relation between the actual probability of fatality associated with a given risk and the value of avoiding that risk.) Despite the great intuitive appeal of this widely held methodological presupposition, I argued that it is simplistic as well as false and that it ought to be replaced with more sophisticated methodological and ethical analyses of the conditions necessary and sufficient for judging a risk to be acceptable. In the next section of this chapter I will briefly describe two alternatives to employment of the linearity assumption. These alternatives are weighted RCBA and use of an adversary process which I call the 'technology tribunal'.

3. TWO NEW METHODS OF RISK ANALYSIS

Obviously no methodological critique of risk assessment is wholly successful if it merely rejects several scientific tools (e.g., the theory of the compensating wage differential, the probability-threshold position, the linearity assumption), but provides no insights for alternative theories, positions, or assumptions to be used in place of those that were rejected. In doing science, as well as risk assessment, one usually has the luxury of abandoning some minimally successful methodological tool only when he has another, and demonstrably better, alternative to it. In this case, I am not prepared to provide a lengthy demonstration of the claim, that my two alternatives to the methods and positions rejected earlier in Chapters 3, 4, 5, and 6 are superior to those against which I argued. I can, however, briefly sketch some of the reasons why I believe such a demonstration easily could be provided. The key to the provision of these reasons lies in the peculiar nature of the epistemological stances likely responsible for the errors criticized in the earlier chapters and hence in the sort of methodological procedures necessary to correct those stances.

One of the most obvious epistemological stances underlying faulty assumptions such as the commensurability presupposition, the linearity assumption, and the probability-threshold position is that there are various algorithms for risk evaluation, and that all the risk assessor need do, in a particular case, is apply the appropriate algorithm. However, the problem with such a hyperrational quest for algorithms, such as the linearity assumption, is that it is doomed to end in oversimplification, and in ignoring the multiplicity and complexity of variables that actually figure in a judgment of risk acceptability. In a recent report on risk assessment, a committee of the U.S. National Academy of Sciences warned of the "cookbook approach to risk assessment," the approach of ignoring the variety of inference options open to an analyst because of both scientific uncertainties and policy judgments.[3] The committee noted that although assessors ought to employ certain methodological guidelines in their analyses, those guidelines ought not to be algorithmic, but flexible; they ought to permit consideration of relevant information unique to a particular risk situation, so as to provide the best judgment for policy decisions.[4]

A second tendency likely responsible for the methodological oversimplifications and errors criticized in the earlier chapters of this volume is assessors' repeated failure to recognize the value components in risk analysis. As was already noted in Section 7 of the previous chapter, those who subscribe to the linearity assumption typically exhibit this tendency. Those who espouse the other positions and methodological assumptions criticized earlier in the volume likewise fall into this same error. In accepting the commensurability presupposition, for example, risk assessors naively assume that the marginal costs of saving lives, across opportunities, ought to be the same, regardless of the fact that risk situations differ – in ethically relevant ways – for example, with respect to the equity of risk distribution, the degree to which the risks are known and compensated, and the extent to which alternative risks provide varying levels of benefits. In simply employing an economic criterion for valuing lives and the costs of saving them, those who subscribe to this presupposition fail to take account of myriad normative components which ought to play a role in evaluation of risks. Likewise, those who accept the probability-threshold position, recently mandated by the U.S. Nuclear Regulatory Commission, also ignore this normative dimension by assuming that a purely scientific criterion, magnitude of probability, determines the acceptability of a risk, apart from whether ethical parameters such as equity, compensation, and benefits have been taken into account.

In failing to recognize the value components in risk assessment, analysts thereby define their enterprise as more objective and scientific than it really is.[5] Moreover, so long as they erroneously believe that their calculated, guestimated risk probabilities are objective (see Section 5 of the previous chapter) and scientific, then they have grounds for arguing that risk analysis is a wholly objective enterprise that ought to be accomplished solely by scientists, rather than a democratic enterprise which ought to be accomplished both by scientists and by citizens affected by the risks. In other words, risk-assessment methodology has implications for social and political theory and practice.

3.1. *An Adversary System of Risk Assessment*

The sorts of methodological errors criticized earlier in this volume arise out of an expert-dominated conception of risk assessment as a wholly objective, purely scientific enterprise. The methodological solutions needed to correct these errors arise out of a cooperative (citizen plus scientist) conception of risk assessment as a normative, policy-oriented enterprise with significant scientific elements. Unless the normative aspect of risk assessment is recognized, and unless the conception of the enterprise is changed accordingly, real negotiation over controversial technological policies will be impossible. This is because the first step in negotiation is mutual recognition of the complex sources of conflict. In this case, the controversy over technological and environmental risk is not merely over scientific methodology, but also over social values. But if this conflict is at least in part a controversy over societal values, and if Thomas Jefferson was correct that the only safe locus of societal power is in the people themselves, then the risk-assessment powers of society ought to be placed in part in the people themselves. If so, then analytic assessors must help both to educate the public and to amend, reformulate, and clarify risk-assessment methods. Such efforts might help to make the real significance of much debate over risk evaluation clearer and it might help to win over alleged opponents of analytic techniques.

In the face of claims that risk assessment has a strong normative and policy dimension, an unfortunate tendency of many persons has been to reject analytic methods and to opt instead for solely political and subjective means of decisionmaking. As I have argued elsewhere, however, this is a mistake; analytic methods of risk assessment are an invaluable tool of public policymaking.[6] Moreover, many celebrated criticisms of analytical risk assessment, for example, those of Amory Lovins, are not based on any in-principle opposition to it, but on disagreement with some particulars of an unrefined method which is still really at its 'basic research' stage.[7] (Lovins' criticisms of risk assessment include difficulties with: determining authentic benefits and risks, unknown and unknowable data, converting risks to costs, and omitting key parameters. He also discusses the limits of risk–benefit analysis, and its inability to address adequately trans-scientific issues, distributive problems, values, and ends, rather than merely means.) Lovins' criticisms appear to be levied against misuse of this method and against insufficient attention to the practical and theoretical problems besetting it. In other words, his complaints are not directed so much at risk assessment, as at those who do not know its place. He warns: "the method's main utility is ancillary. It encourages specificity in stating the grounds on which a project is believed to be a good or bad idea and rigor in approaching the genuine limits of quantification .... These benefits can be obtained without deifying the results of analysis ...."[8] The tenor of Lovins' remarks suggests that he could well be amenable to proper use of a more sophisticated version of risk assessment. If so, then rational debate, both about risk analysis and about issues of great substantive disagreement, might be possible. A necessary condition for such rational debate, of course, is that persons accept the in-principle desirability of some analytic tools of risk assessment and that they be willing to discuss what sorts of modifications in existing analytic methods are most acceptable. In other words, once people are willing, in principle, to use some analytic method as a policymaking instrument, then real progress in improving risk assessment is possible. Given this common belief in rational analysis as a policy tool, persons as diverse as Lovins and Rasmussen might be able to make logical contact with each other on the basis of an accepted risk-assessment methodology. As it is now, too much of the talk is at cross-purposes, and too much of the debate is between those who ignore values and those who simply assume that their particular evaluative assumptions are beyond question.

Much risk-assessment talk is at cross-purposes because many experts are emphasizing the wrong issue. The issue is not only how to make the public more 'rational', more accepting of the linearity assumption or the commensurability presupposition, for example. Rather, the real issue is also how to accommodate democratic values within analytic assessment. It is surely not appropriate, when citizens exhibit a high aversion to low-probability, uncompensated, involuntarily imposed risk, to miss the *ethical* import of such reluctance, and to describe this aversion as an 'irrational' exhibition of a discontinuous preference function.[9] Instead, assessors need to seek ways to represent important *values* within analytic assessment. Admittedly the assessment cannot and ought not produce those values. The most we can ask of a good analytical technique is to clarify our logical, scientific, and ethical options. It is not, and ought not be, a substitute for the dynamics of the political system.

But if the arguments made earlier in this volume are correct, then adherence to the commensurability presupposition, the theory of the compensating-wage differential, the probability-threshold position, the linearity assumption, and other methodological tools of risk assessors does not contribute to the goal of clarifying our logical, scientific, and ethical options regarding societal risks. (And, as was just argued, perhaps this failure has occurred in part because adherence to these doctrines is premissed both on a misguided search for algorithms able to deliver up correct risk evaluations and on ignoring the normative components of risk analysis.) But if some of the classical methods of risk assessment do not contribute to this goal, then the obvious question is whether there are other methods and techniques which are likely to be more fruitful. I believe that there are.

Pursuing the insight that several current methods of risk assessment have failed because analysts ignored the value components in their work, I believe that any fruitful method of risk analysis must explicitly address controversies over values. One of the best ways to do this is to pursue an adversary method of assessment, a method premised on the fact that desirable risk analyses are likely to be a product of rational interaction and compromise among those who disagree about how to evaluate a given risk.

Although the risk-assessment committee of the National Academy of Sciences has neither explicitly proposed nor endorsed the adversary method to which I subscribe, it is important to note that the committee does support several methodological principles which are the foundation of adversary assessment. Some of these principles include: establishment of a clear distinction between scientific and policy judgments; making draft risk assessments available to the public before regulatory action is taken; and requiring all risk assessments to be reviewed by independent advisory panels before they are used as the basis of public policy.[10] Likewise, outside the U.S. there also appears to be support, in principle, for adversary methods of risk evaluation. At a recent meeting on risk assessment, sponsored by the Commission of the European Communities, experts concluded that the risk-assessment process needed to include serious involvement of "nontechnical people". The group noted that such involvement would likely increase conflict over particular risks and their assessment, and that this conflict would likely slow the decisionmaking process. They affirmed, however, that including representative citizens, in an adversary setting, would "enrich the debate" and widen the perspective on the basis of which ultimate risk decisions were made. Moreover, they argued that public participation, in an adversary setting, is unavoidable anyway, because most serious risk situations have already become politicized.[11] Because of the value-laden nature of risk assessment, expert participants in the conference sponsored by the Commission of the European Communities concluded that there was a "need for new decisionmaking mechanisms" and that such mechanisms ought to be both collaborative (thus relying on the best scientific judgment available) and adversarial.[12]

Since I have discussed, elsewhere, my own proposal for an adversary method of risk assessment,[13] I will not repeat my analysis here. I can, however, provide a brief overview of the adversary methods for which I have argued. Consisting of three distinct stages, I call my proposal 'the technology tribunal'. The first stage provides for the establishment of a tribunal, composed of scientists and citizens, to identify the significant questions of science, technology, and policy associated with the controversial issue in question. The second stage is an adversary proceeding presided over by a panel of impartial scientists and laymen, none of whom have conflicts of interest regarding the issue at hand. During this proceeding, advocates debate the technical and policy questions that are in dispute. In addition to presenting their own cases, the debaters are able to cross-examine opponents and to criticize their arguments. At the third and final stage of the court procedure, the panel of judges issues its decision as to the scientific and policy factors relevant to the disputed questions. This decision is made public, unless national security dictates otherwise, and is designed to provide the basis for reaching political decisions through the democratic process.

Somewhat similar to the 'Science Court' proposed in 1978 by the Carter Administration, the technology tribunal differs in two major respects from the earlier proposal. First, it consists of a panel of scientists and educated laymen, rather than only of a panel of scientists, as was suggested for the science court. Second, also contrary to the science-court proposal, there is no distinction between facts and values, no attempt to have the tribunal treat only the scientific facts of the matter and set aside the policy or value issues. Rather, the technology tribunal is designed to deal with both sorts of issues; this is because, epistemologically, facts and values cannot be separated and politically, risk controversies are usually neither wholly factual nor wholly evaluative. Given that these two dimensions of the conflict cannot be separated (a point for which I have argued elsewhere), and that every risk decision has important health and safety consequences for the public, I maintain that citizens, informed by scientists' participation in the proceedings, ought to join the experts in helping to evaluate risks and to provide a democratic basis for subsequent public policymaking.[14] My basic argument is that, just as Congresspeople and Presidents can be informed adequately on a science-related issue, so as to deal with it adequately, even though they are not themselves scientists, so also educated laymen ought to be able to do the same thing, and hence to inform risk assessment with democratic values. I maintain that not only is citizen participation possible, it is also desirable, since citizens have a right to informed consent over the conditions, like risks, affecting their well-being. Given the in-principle plausibility of the technology tribunal, I hold that the number of panelists serving on the tribunal could be anywhere from several dozen to several hundred or several thousand, depending on the nature of the issue and the desires of the public.

Although I have argued elsewhere for the technology tribunal,[15] it is important to note that my precise plea is for an "experiment" to this effect, complete with citizen adjudication, as has already been tried on a small scale in cities such as Ann Arbor, Michigan and Cambridge, Massachusetts. Although I am a believer in the importance of this experiment, in part because of some of the problems spelled out earlier in this volume, I am an agnostic regarding its ultimate success. I am confident that we as a society at least should attempt to examine the political, educational, and scientific consequences of adversary proceedings carried out in democratic, rather than elitist, fashion. Moreover, the very debate engendered by the experiment, including equal funding for advocates to prepare alternative risk assessments, could provide sufficient enlightenment to justify the expense of time and energy in the experiment. If we as a society could admit, at the outset, that alternative and controversial stances on various risk-assessment methods, assumptions, and conclusions were possible, and that the normative components of these analyses ought to be rendered explicit and subject to democratic debate as soon as possible in the assessment proceedings, then we might be willing to attempt the technology-tribunal experiment.

3.2. *Weighted Risk Assessments*

The methodological flaws covered earlier in this volume, however, arise not only because assessors ignore the normative dimensions of their task, but also because, in a scientific sense, they oversimplify it. Just as the technology tribunal is one way to deal with the unrecognized evaluative issues in a risk assessment, so also there is a way to deal with the problem of scientific oversimplification.

In the most obvious sense, what went wrong with the commensurability presupposition, the compensating wage differential, the probability-threshold position, and the linearity assumption (criticized earlier in this volume) is that assessors attempted to judge the relative value of avoiding various risks by using too simple a mathematical model.

The same type of mistake, an error of oversimplification, has often been made in energy modelling. A number of authors have assumed, for example, that increases in the GNP are proportional to increases in energy consumption,[16] just as many risk assessors have assumed, for example, that risk aversion is proportional to the probability of fatality of the risk. Just as the risk assessors might need a multivariate mathematical function incorporating parameters such as the risk's degree of voluntariness, if they are to evaluate societal risks, so also energy theorists have discovered that they need a multivariate function incorporating parameters such as the degree of industrialization, the size of the service sector, and the energy-to-output ratio, if they are to evaluate energy needs for a given area.[17]

The problem, of course, is that there are no multivariate functions 'around the corner', nor are we likely to discover any, whether to assess energy needs or technological risks. An alternative, both to the crude attempt to describe risk in terms of the commensurability presupposition or the linearity assumption, or as a function merely of probability of fatality, and to the impossible task of describing it according to some multivariate function, might be to use a modified expected-value or expected-utility model, in conjunction with some alternative ethical assessment theories. What might this new model be like? As I have argued elsewhere (see note 30), it could be based on aggregating utilities, as is the cost–benefit model, but it might include factors ordinarily ignored in classical cost–benefit analysis, factors such as the effects of incomplete information or distributive inequity on utility.

By using factors such as these to *weight* the various risks, costs, and benefits, one might help to resolve a classic problem of utility theory, namely, the difficulty that low-probability/high-consequence events and high-probability/low-consequence events could have the same expected value.[18] High-consequence situations could be weighted more heavily, for example, on the basis of parameters such as catastrophic potential or societal disruption. One might also use a hierarchical ordering of alternative ethical rules as a way to weight or to structure a class of decisions, rather than to use a cardinal utility framework.[19] Such an ordering could have the merit of making risk assessment amenable to a number of different ethical frameworks, not just utilitarianism.

A good way to begin to devise such a modified expected-value model might be to determine all the types of parameters (e.g., probability of risk, its equity of distribution, possible compensation) allegedly relevant to risk evaluation. After all these factors were enumerated, society could decide on a number of ethical assessment schemes (e.g., utilitarianism, egalitarianism) according to which the factors would be weighted.[20] The ethical assessment schemes could be formulated in terms of transitive, hierarchical lists of ethical rules according to which the relative values of the risk factors were weighted.

In a Rawlsian framework, one such rule might take the form of attempting to maximize the utility of the worst-off person in society, so long as he remained the worse off.[21] Likewise, in the Paretian framework, one rule might take the form of attempting to maximize the utility of one or more persons, so long as no one were harmed; another such rule might require gainers from a particular safety program to compensate losers.[22] By means of using such rules, one could thus examine alternative, ethically weighted options, and not merely probability of fatality of marginal cost across opportunities.

Admittedly, however, such a modified expected-value model has a number of limitations. Although it might enable risk analysis to escape purely utilitarian values, hierarchical, transitive, ethical rules which were simple and clear enough to follow might be alleged by some either to represent a particular ethical position inadequately or to fail to cover certain circumstances. Other persons might claim that the ethical rules were too complicated and academic, or too difficult to apply. There would also be the problem of deciding who would set the relative weights or determine which hierarchically ordered ethical rules would be followed. Should the weighting scheme reflect democratic, meritocratic or autocratic values?

While these are reasonable criticisms, none of them seems to me to address any in-principle difficulty with an ethically expanded expected value/utility model. Admittedly, such a model would be helpful primarily in *clarifying* risk situations as they are affected by value parameters, such as the equity of risk distribution. The model would not likely be successful in predicting societal risk behavior in situations where decisions were made on the basis of limited knowledge. As Kunreuther has suggested, however, one might modify the expected-utility model so as to incorporate the costs of making decisions, i.e., translate the efforts of gathering and processing information into assessment terms. Thus limited human knowledge and experience might be considered as an explicit part of the utility function. As a consequence, the function might come closer to providing insights into events not currently capable of being treated in classical expected-utility theory, events such as individuals' failure to: (a) buy flood insurance, even though it is 90% subsidized; (b) wear seat belts, and (c) purchase crime insurance, even though it is highly subsidized.[23]

Consider, too, that such a lexicographic or hierarchical weighting scheme may be no more difficult to accomplish, econometrically, legally, and politically, than Mishan's proposed assimilation of externalities within the cost–benefit framework. Yet, such an assimiliation is widely acknowledged to be absolutely essential to accurate risk analysis. Moreover, although economists have not been using ethically weighted schemes in the precise ways just suggested, they have traditionally used a discounting procedure in an attempt to make future values commensurate with present ones.

Economists usually justify the use of a discount rate on future benefits and costs (i.e., weighting them at less than current economic values) because a given amount of money invested now will yield a much larger total because of the rate of return over inflation. For example, in 50 years, if we were to give a "fair" compensation of $2,500 to some future individual for effects of a carcinogen we are currently using, then we would only need to invest $558 now, if there were a 3% real rate of return over inflation in the next 50 years.[24] Of course, there are numerous controversies surrounding both the justification for, and the application of, these traditional discounting schemes, especially as regards future public good.[25] I do not wish to argue about their merit here. My point is simply that such techniques provide a useful model for the sort of weighting which might be possible within an analytic framework. Just as, rightly or wrongly, current economic theory allows one to weight differently the utilities of different persons and generations, so also a modified method of risk assessment might allow us to use principles of discounting to weight differently risks which are equitable or inequitable, private or public, compensated or uncompensated, reversible or irreversible, or which lead to one type of effects as opposed to another type.[26]

Ben-David, Kneese, and Schulze (1979) showed recently that parameters of the same risk assessment could be given varying weights on the basis of alternative ethical criteria. By means of following four respective ethical systems, each of which was expressed in terms of different, transitive, hierarchical rules, the analysis could be shown to justify different conclusions about net cost-effectiveness. They illustrated, for example, that when Benthamite weighting criteria were used, the costs of nuclear power could be shown to outweigh the benefits, but that when traditional discounting procedures were employed, the benefits could be said to exceed the costs.[27] Likewise, they illustrated that when the costs and benefits of the automotive emission control standards (as laid down by the amendment to the 1970 Clean Air Act) were weighted according to Rawlsian ethics, they could be shown to be feasible and economical. When these same costs and benefits were weighted according to Benthamite ethics, however, they could be shown to be unfeasible.[28] If one could weight risk parameters, either on the basis of various ethical criteria, or on the basis of inputs from, or alternative assessments done by, various interest groups,[29] then much the same results would likely follow in various risk analyses as in the Ben-David, Schulze, and Kneese study. If one weighted various probabilities and magnitudes, including factors such as equity, along some of the lines discussed, instead of assuming that the value of risk avoidance was a simple function of the probability of fatality associated with the risk, then risk assessors might move one step closer to a more comprehensive risk model, to understanding public perception of risk, and to closing the gap between expert and lay perceptions of risk.[30]

3.3. *The Future of Risk Assessment*

It may be, however, that such an ethically weighted risk assessment method will not work. As with the technology tribunal, I am an agnostic as to the ultimate success of the weighting scheme mentioned here. As to the necessity of risk assessors' performing a weighting-scheme experiment, however, I am a believer.

Although neither the technology tribunal nor weighted risk assessment can provide any ethical or methodological insights about how to evaluate societal risks, they should go at least part of the way toward giving us frameworks which allow us to represent values (equity of distribution, compensation, voluntariness) often previously ignored by risk assessors. However deficient, both proposals provide for explicit, systematic formulation of various risk parameters and for alternative ways of weighting or valuing them. Because they are systematic, they are amenable both to being understood by the public and to democratic modification.

Opponents of analytic risk assessment, amended by means of these two proposals, must bear the burden of defending some other method of public policymaking, if they are unwilling to consider the two experiments proposed here. What is certain is that we cannot continue the *status-quo* acceptance of doctrines like the commensurability presupposition or the linearity assumption. We must improve our methods for dealing with societal risk. Given no philosopher king and no benevolent scientist-dictator, our task will be difficult.

NOTES

[1] This suggestion is also made by D. M. Berman, *Death on the Job*, Monthly Review Press, New York, 1978, p. 192; hereafter cited as: Berman, *DOJ*.

[2] Berman, *DOJ*, pp. 191–192, also makes these suggestions.

[3] See Frank Press, *Risk Assessment in the Federal Government: Managing the Process*, National Academy Press, Washington, D.C., 1983, p. 74; hereafter cited as *RA*.
[4] Press, *RA*, pp. 163–165.
[5] A recent report of a committee of the National Academy of Sciences noted emphatically that risk assessment was not merely a neutral, scientific enterprise, but that it had policy, as well as scientific components. See Press, *RA*, pp. 48–49.
[6] For arguments in support of analytic risk assessment, see K. S. Shrader-Frechette, *Science Policy, Ethics, and Economic Methodology*, Reidel, Boston, 1984, Chapter Two.
[7] See A. Lovins, 'Cost–Risk–Benefit Assessment in Energy Policy', *George Washington Law Review* **45** (5), (1977), 911–943; hereafter cited as: CRBA.
[8] Lovins, CRBA, p. 942.
[9] See S. Burns, 'Congress and the Office of Technology Assessment', *George Washington Law Review* **45** (5), (1977), 1150; hereafter cited as: Congress. See also Lovins, CRBA, p. 929.
[10] Press, *RA*, pp. 151–157.
[11] See R. Coppock, 'Discussion and Suggestions for Further Action', in M. Dierkes, S. Edwards, and R. Coppock (eds.), *Technological Risk*, Verlag Anton Hain, Königstein, 1980, pp. 133–134; hereafter cited as: Coppock, Suggestions.
[12] Coppock, Suggestions, pp. 134–135.
[13] See K. S. Shrader-Frechette, *Science Policy, Ethics, and Economic Methodology*, Reidel, Boston, 1984, Chapter Nine; hereafter cited as: Science Policy.
[14] See Shrader-Frechette, Science Policy, Chapter Nine.
[15] See Shrader-Frechette, Science Policy, Chapter Nine.
[16] See, for example, Häfele, Energy, pp. 136–137, and M. Willrich, *Global Politics of Nuclear Energy*, Praeger, New York, 1971, pp. 184–186.
[17] See, for example, Robert Stobaugh, 'After the Peak: The Threat of Imported Oil'; Mel Horwitch, 'Coal: Constrained Abundance'; and Daniel Yergin, 'Conservation', in Robert Stobaugh and Daniel Yergin (eds.), *Energy Future*, Random House, New York, 1979, pp. 37; 84–85; 142–143; 155, respectively.
[18] H. Kunreuther, 'Limited Knowledge and Insurance Protection', in D. Okrent, (ed.) *Risk–Benefit Methodology and Application*, UCLA School of Engineering and Applied Science, Los Angeles, 1975, pp. 175–198; hereafter cited as: LK.
[19] Kunreuther, LK, p. 195.
[20] S. Ben-David, A. V. Kneese, and W. D. Schulze, *A Study of the Ethical Foundations of Benefit-Cost Analysis Techniques*. Working paper. Done under

NSF-EVIST funding. August 1979, pp. 11–22. Hereafter cited as: Ben-David, Kneese, and Schulze, *BCA*.

21 J. Rawls, *A Theory of Justice*, Harvard University Press, Cambridge, 1971.

22 See Ben-David, Kneese, and Schulze, *BCA*, pp. 16–21.

23 Kunreuther, LK, pp. 192, 175.

24 Ben-David, Kneese, and Schulze, *BCA*, pp. 8–9.

25 See Ben-David, Kneese, and Schulze, *BCA*, pp. 32 ff.

26 See Rowe, *Risk*, pp. 312, 344, and A. Porter, F. Rossini, S. Carpenter, and A. Roper, *A Guidebook for Technology Assessment and Environmental Impact Analysis*, North Holland, New York, 1980, pp. 270–272.

27 Ben-David, Kneese, and Schulze, *BCA*, pp. 98, 104–105.

28 Ben-David, Kneese, and Schulze, *BCA*, pp. 121–130.

29 See Green, CBA, pp. 287–288 and W. Häfele, 'Benefit–Risk Tradeoffs . . .', in H. Ashley, R. Rudman, and C. Whipple (eds.), *Energy and the Environment* Pergamon, New York, 1976, p. 179; hereafter cited as: BRT.

30 For a discussion of an ethical weighting scheme for risk assessment, see K. S. Shrader-Frechette, Science Policy, Chapter Eight.

SELECTED BIBLIOGRAPHY
RISK ASSESSMENT

Aharoni, Yair: 1981, *The No-Risk Society*, Chatham Press, Old Greenwich, Ct.

Arrow, K. J.: 1964, *Social Choice and Individual Values*, 2nd ed., Wiley, New York.

Ascher, William: 1979, 'Problems of Forecasting and Technology Assessment', *Technological Forecasting and Social Change* **13**, 150–151.

Ashley, H., *et al.* (eds.): 1976, *Energy and the Environment: A Risk–Benefit Approach*, Pergamon, New York.

Atomic Energy Commission: 1974, *Comparative Risk–Cost–Benefit Study of Alternative Sources of Electrical Energy*, Report WASH-1224, U.S. Atomic Energy Commission, Washington, D.C.

Baker, R. F., Michaels, R. M., and Preston, E. S.: 1975, *Public Policy Development: Linking the Technical and Political Processes*, John Wiley and Sons, New York.

Baram, Michael S.: 1980, 'Cost–Benefit Analysis: An Inadequate Basis for Health, Safety, and Environmental Regulatory Decisionmaking', *Ecology Law Quarterly* **8**, 477–526.

Barnthouse, L. W., *et al.*: 1982, *Methodology for Environmental Risk Analysis*. Oakridge National Laboratory Report ORNL/TM-8167, National Technical Information Service, Springfield, Virginia.

Bazelon, D.: 1979, 'Risk and Responsibility', *Science* **205**, 277–280.

Berg, G. G. and Maillie, H. D. (eds.): 1981, *Measurement of Risks*, Plenum Press, New York.

Bierman, H., Jr., Bonini, C. P., and Hausman, W. H.: 1973, *Quantitative Analysis for Business Decisions*, Richard D. Irwin, Inc., Homewood, Illinois.

Brown, Bowden V.: 1976, 'Projected Environmental Harm: Judicial Acceptance of a Concept of Uncertain Risk', *Journal of Urban Law*, February, 525–529.

Burns, S.: 1977, 'Congress and the Office of Technology Assessment', *George Washington Law Review* **45**, 1123–1150.

Burton, I., Fowle, C. D., and McCullough, R. S. (eds.): 1982, *Living with Risk: Environmental Risk Management in Canada*, Institute for Environmental Studies, University of Toronto, Canada.

Cairns, John, Jr., Dickson, K. L., and Maki, A. W. (eds.): 1978, *Estimating the Hazard of Chemical Substances to Aquatic Life*, ASTM STP 657, American Society for Testing and Materials, Philadelphia, Pennsylvania.

Caputo, R.: 1977, 'An Initial Comparative Assessment of Orbital and Terrestrial Center Power Systems', Jet Propulsion Laboratory, Report 900-780, Pasadena, California.

Caputo, R.: 1979, Letter, *Science* **204**, 454.

Carley, Michael J. and Derow, E. O.: 1980, *Social Impact Assessment: A Cross-Disciplinary Guide to the Literature*, Research Paper 80–1, Policy Studies Institute, London.

Carter, Luther J.: 1979, 'Dispute Over Risk Quantification', *Science* **203**, 1324–1325.

Clark, E. M. and Van Horn, A. J.: 1976, 'Risk–Benefit Analysis and Public Policy: A Bibliography', Nov. 1976. Updated and extended by L. Hedal and E. A. C. Crouch, Energy and Environmental Policy Center, Harvard University, Cambridge, Massachusetts, 1978.

Coates, Joseph: 1974, 'Some Methods and Techniques for Comprehensive Impact Assessment', *Technological Forecasting and Social Change* **6**, 341–357.

Cohen, B. L.: 1981, 'Long Term Consequences of the Linear-No-Threshold Dose-Response Relationship for Chemical Carcinogens', *Risk Analysis* **1**, 267–275.

Cohen, B. L. and Sing Lee, I.: 1979, 'A Catalog of Risks', *Health Physics* **36**, 707–722.

Comar, C.: 1979, 'Risk: A Pragmatic *De Minimus* Approach', *Science* **203**, 319.

Committee on the Institutional Means for Assessment of Risks to Public Health (Reuel A. Stallones, Chairman), Commission on Life Sciences, National Research Council: 1983, *Risk Assessment in the Federal Government: Managing the Process*, National Academy Press, Washington, D.C.

Committee on Public Engineering Policy (ed.): 1972, *Perspectives on Benefit–Risk Decision Making*, National Academy of Engineering, Washington, D.C.

Committee on Risk and Decisionmaking (Howard Raiffa, Chairman), Assembly of Behavioral and Social Sciences, National Research Council: 1982, *Risk and Decisionmaking: Perceptions and Research*, National Academy Press, Washington, D.C.

Conrad, J. (ed.): 1980, *Society, Technology, and Risk Assessment*, Academic Press, London.

Conway, R. A. (ed.): 1982, *Environmental Risk Analysis for Chemicals*, Van Nostrand Reinhold Company, New York.

Coppola, A. and Hall, R. E.: 1981, *A Risk Comparison* (NUREG/CR-1916), Brookhaven National Laboratory, prepared for the Nuclear Regulatory Commission, Upton, New York.

Covello, V. T. and Abernathy, M.: 1982, 'Risk Analysis and Technological Hazards: A Policy-Related Bibliography', National Science Foundation, draft report.

Covello, V. T., Menkes, J., and Nehnevajsa, J.: 1982, 'Risk Analysis, Philosophy, and the Social and Behavioral Sciences: Reflections on the Scope of Risk Analysis Research', *Risk Analysis* **2**, 53–58.

Crandall, Robert W., and Lave, L. B.: 1981, *The Scientific Basis of Health and Safety Regulation*, The Brookings Institution, Washington, D.C.

Crouch, E. A. C. and Wilson, R.: 1982, *Risk/Benefit Analysis*, Ballinger Publishing Company, Cambridge, Massachusetts.

Cumming, R. B.: 1981, 'The Urge To Shout', *Risk Analysis* **1**, 165.

Daddario, Emilio, Q., quoted by R. A. Carpenter in *Technology Assessment*, 1970, Hearings before the Subcommittee on Science, Research and Development, Committee on Science and Astronautics, U.S. House of Representatives, 91st Congress (1st Session, 1969 [No. 13]) U.S.G.P.O.

Derr, Patrick, Goble, R., Kasperson, R. E., and Kates, R. N.: 1981, 'Worker/Public Protection: The Double Standard', *Environment* **23**, 15, 35.

Douglas, M. and Wildavsky, A.: 1982, *Risk and Culture*, University of California Press, Berkeley.

Dierkes, M., Edwards, S., and Coppock, R.: 1980, *Technological Risk*, Oelgeschlager, Gunn, and Hain, Cambridge.

Etzioni, Amitai: 1979, 'How Much Is a Life Worth?', *Social Policy*, March/April, 6.

Farmer, F. R.: 1980, 'Methodology of Energy Risk Comparisons', *Intl. Atomic Energy Agency Bull.* **22**, 120.

Finsterbush, K. and Wolf, C. P.: 1977, *The Methodology of Social Impact Assessment*, Dowden, Hutchinson, and Ross Publishing Co., Stroudsberg, Pennsylvania.

Fischhoff, B., Lichtenstein, S., Slovic, P., Derby, S., and Keeney, R.: 1981, *Acceptable Risk: A Critical Guide*, Cambridge University Press, New Rochelle, New York.

Fischhoff, B., *et al.*: 1978, 'How Safe Is Safe Enough?', *Policy Sciences* **9**, 127–152.

Fischhoff, B., *et al.*: 1979, 'Weighing the Risks', *Environment* **21**, 17–20, 32–38.

Fowles, Jib: 1978, *Handbook of Futures Research*, Greenwood Press, Westport, Connecticut.

Graham, J. D. and Vaupel, J. W., 1981, 'Value of Life: What Difference Does It Make?' *Risk Analysis* **1**, 89–91.

Gusman, S., Moltke, K. W., Irwin, F., and Whitehead, C.: 1980, *Public Policy for Chemicals – National and International Issues*, The Conservation Foundation, Washington, D.C.

Hamburg, Morris: 1977, *Statistical Analysis for Decision-Making*, 2nd Edition, Harcourt Brace Jovanovich, New York.

Hamilton, L. D.: 1980, 'Comparative Risks from Different Energy Systems: Evolution of the Methods of Studies', *Intl. Atomic Energy Agency Bull.* **22**, 35.

Harris, Louis and Associates: 1980, *Risk in a Complex Society: A Marsh and McLennan Public Opinion Survey*, Marsh and McLennan, New York.

Herbert, J., Swanson, C., and Reddy, P.: 1979, 'A Risky Business', *Environment*, July–August, 28–33.

Herrera, G. (ed.): 1977, 'Assessment of R, D & D Resources, Health and Environmental Effects, O & M Costs and Other Social Costs for Conventional and Terrestrial Solar Electric Plants', Jet Propulsion Laboratory, Report 900-782, Pasadena, California.

Hertz, D. B. and Thomas, H.: 1982, *Risk Analysis*, Wiley, New York.

Hetman, F.: 1973, *Society and the Assessment of Technology*, Organization for Economic Cooperation and Development, Paris.

Hohenemser, C. and Kasperson, J.: 1982, *Risk in the Technological Society*, Westview, Boulder.

Hohenemser, C., Kates, R. W., and Slovic, P.: 1983, 'The Nature of Technological Hazard', *Science* **220**, 378–384.

Holden, Constance: 1980, 'Love Canal Residents Under Stress', *Science* **208**, 1242.

Holdren, J. P., Smith, K., and Morris, G.: 1979, Letter, *Science* **204**, 564–567.

Holdren, J.: 1981, 'Risk of Energy Options', *Risk Analysis* **1**, 41.

Holdren, J. P., Anderson, K., Gleick, P. H., Mintzer, I., Morris, G., and Smith, K. R.: 1979, 'Risk of Renewable Energy Sources: A Critique of the Inhaber Report', Energy and Resources Group, Report ERG 79-3, University of California, Berkeley, California.

Holdren, J. P.: 1982, 'Science and Personalities Revisited', *Risk Analysis* **1**, 173–176.

Hoos, Ida R.: 1979, 'Societal Aspects of Technology Assessment', *Technological Forecasting and Social Change* **13**, 191–202.

Horowitz, T. L. and Katz, J. E.: 1975, *Social Science and Public Policy in the United States*, Praeger Publications, New York.

House, P. W.: 1981, *Comparing Energy Technology Alternatives from an Environmental Perspective*, U.S. Department of Energy, Washington, D.C., DOE/EV-0109.

House, P. W.: 1977, *Trading Off Environment, Economics, and Energy: A Case Study of EPA's Strategic Environmental Assessment System*, D. C. Heath and Company, New York.

Huddle, Franklin, P., Science Policy Research Division, Congressional Research Service, Library of Congress: 1972, *Science Policy: A Working Glossary*. Prepared for the Subcommittee on Science, Research and Development, Committee on Science and Astronautics, U.S. House of Representatives, 92nd Congress. U.S. Government Printing Office, Washington, D.C.

Hull, A. P.: 1982, 'The Limits of the Peer Review Process', *Risk Analysis* **1**, 177–178.

Inhaber, H.: 1981, 'Energy Production Risks: Science and Personalities', *Risk Analysis* **1**, 9.

Inhaber, H.: 1983, *Energy Risk Assessment*, Gordon and Breach, New York.

Inhaber, H.: 1978, *Risk of Energy Production*, Atomic Energy Control Board Report AECB-1119, Ottawa, Canada, March 1978; May 1978; Nov. 1978. See also: *New Scientist* **78**, 444–446 and *Energy, the International Journal* **3**, 778–789.

Inhaber, H.: 1979, 'Risk with Energy from Conventional and Nonconventional Sources', *Science* **203**, 718–723.

Jones, Martin: 1971, *A Technology Assessment Methodology: Basic Propositions*, MITRE Corporation, McLean, Virginia.

Kates, R. W.: 1975, 'Comparative Risk Assessment', a Report of a Workshop on Environmental Hazards in an International Context, Woods Hole, MA, March 31–April 4, 1975, for Scientific Committee on Problems of the Environment, International Council of Scientific Unions.

Kates, R. W.: 1978, *Risk Assessment of Environmental Hazard*, SCOPE Report No. 8., John Wiley & Sons, New York.

Keeney, Ralph L. and Raiffa, H.: 1976, *Decisions with Multiple Objectives – Preferences and Value Tradeoffs*, John Wiley & Sons, New York.

King, W. R. and Cleland, D. I.: 1978, *Strategic Planning and Policy*, Von Nostrand Reinhold, New York.

Kletz, Trevor A.: 1977, 'The Risk Equation: What Risks Should We Run?', *New Scientist*, May 12, 322.

Kunreuther, H. and Ley, E.: 1982, *The Risk Analysis Controversy*, Springer Verlag, New York.

Kunreuther, H. J. and Linnerooth, J. (eds.): 1983, *Risk Analysis and Decision Processes*, Springer-Verlag, New York.

Lapp, Ralph: 1981, 'Cancer and the Fear of Radiation', *New Scientist*, July 2, 15.

Lave, Lester B.: 1981, *Economic Implications of Shifting from Federal Regulation to Its Alternatives*, unpublished paper, Brookings Institution, Washington, D.C.

Lave, Lester B. (ed.): 1982, *Quantitative Risk Assessment in Regulation*, Brookings Institution, Washington, D.C.

Lave, Lester B.: 1981, *The Strategy of Social Regulation*, Brookings Institution, Washington, D.C.

Lawless, Edward, Jones, M., and Jones, R. M.: 1983, *Comparative Risk Assessment*, Final Report, NSF Grant No. PRA-8018868, National Science Foundation, Washington, D.C.

Lemberg, Rein: 1979, 'Energy: Calculating the Risks', *Science* **204**, 454.

Linstone, Harold A., *et al.*: 1978, *The Use of Structural Modeling for Technology Assessment*, 2 Vols., Portland State University, Portland, Oregon.

Lovins, Amory B.: 1977, 'Cost–Risk–Benefit Assessment in Energy Policy', *The George Washington Law Review* **45**, 911–943.

Lowrance W. W.: 1976, *Of Acceptable Risk: Science and the Determination of Safety*, William Kaufman, Los Angeles, California.

Luce, R. D. and Raiffa, H.: 1975, *Games and Decisions*, John Wiley & Sons, New York.

Mishan, E. J.: 1976, *Cost–Benefit Analysis*, Praeger, New York.

Mitchell, Arnold, *et al.*: 1975, *Handbook of Forecasting Techniques*. Stanford Research Institute, Palo Alto, Report on Contract DACW 31-75-C-0027. United States Army Corps of Engineers. IWR Report No. 75–7.

Mitre Corporation: 1979, *Symposium/Workshop . . . Risk Assessment and Governmental Decision Making*, The Mitre Corporation, McLean, Virginia.

Moll, K. D., *et al.*: 1975, 'Hazardous Wastes: A Risk–Benefit Framework Applied to Cadmium and Asbestos', SRI International, Menlo Park, CA.

Moll, K. D. and Tihansky, D. P.: 1977, 'Risk–Benefit Analysis for Industrial and Social Needs' (a summary paper emphasizing asbestos), *American Industrial Hygiene Association, J.* **38**, 153–161.

Moss, T. and Lubin, B.: 1981, 'Risk Analysis: A Legislative Perspective', in *Health Risk Analysis* (ed. by C. R. Richmond, P. J. Welsh, and E. O. Copenhauer), Frankin Institute Press, Philadelphia, pp. 29–50.

Naylor, T. H. and Schauland, H. A.: 1976, 'A Survey of Users of Corporate Planning', *Management Science* **22**, 927–937.

Nemetz, Peter N. and Vining, A. R.: 1981, 'The Biology–Policy Interface: Theories of Pathogenesis, Benefit Valuation and Public Policy Formation', *Policy Sciences* **13**, 35, 127.

Nicholson, W. J. (ed.): 1981, *Management of Assessed Risk for Carcinogens, Annals of New York Academy of Sciences*, Vol. 63.

Office of Technology Assessment, U.S. Congress: 1981, *Assessment of Technologies for Determining Cancer Risks from the Environment*, U.S. Government Printing Office, Washington, D.C.

Office of Technology Assessment, U.S. Congress: 1979, 'Methods for Assessing Health Risks', Chapter V (pp. 59–70) in *Environmental Contaminants in Foods*. OTA-F-103; U.S. GPO (Stock No. 052-003-00724-0), Washington, D.C.

Okrent, D. (ed.): 1975, *Risk–Benefit Methodology and Application* (UCLA-ENG-7598), University of California, Los Angeles.

Okrent, D.: 1980, 'Comment on Societal Risk', *Science* **208**, 372–375.

Okrent, D. and Whipple, C.: 1977, *Approach to Societal Risk Acceptance Criteria and Risk Management*. PB-271 264, U.S. Department of Commerce, Washington, D.C.

Ott, W. R.: 1978, *Environmental Indices – Theory and Practice*, Ann Arbor Science Publishers, Inc., Ann Arbor, Michigan.

Otway, H. J. and Pahner, P. D.: 1976, 'Risk Assessment', *Futures* **8**, 122–123.

Otway, H.: 1977, 'Risk Assessment and the Social Response to Nuclear Power', *Journal of the British Nuclear Engineering Society* **16**, 327–333.

Otway, H. J.: 1971, *Risk vs. Benefit: Solution or Dream?* Compendium of papers presented at symposium sponsored by Western Interstate Nuclear Board, Los Alamos Scientific Laboratory, Los Alamos, New Mexico. November 11–12.

Parzyck, D. C. and Inhaber, H.: 1980, 'Comparative Health Risks of Energy Production', Oak Ridge National Lab., TN.

Petak, W. and Atkisson, A.: 1982, *Natural Hazard Risk Assessment and Public Policy*, Springer-Verlag, New York.

Porter, Alan L., *et al.*: 1980, *A Guidebook for Technology Assessment and Impact Analysis*, North Holland, New York.

Pouliquen, L. Y.: 1970, *Risk Analysis in Project Appraisal*, Johns Hopkins, Baltimore.

Press, F., Chair, Committee on the . . . Assessment of Risks to Public Health, National Research Council: 1983, *Risk Assessment in the Federal Government*, National Academy Press, Washington, D.C.

Raiffa, H.: 1968, *Decision Analysis: Introductory Lectures on Choice Under Uncertainty*, Addison-Wesley, Reading, Massachusetts.

Rau, J. G. and Wooten, D. C.: 1980, *Environmental Impact Analysis Handbook*, McGraw-Hill Company, New York.

Reissland, J. and Harries, V.: 1979, 'A Scale for Mearuring Risks', *New Scientist* **83**, 809–811.

Rejda, G. and Ginsberg, R.: 1979, *Risk and Its Treatment*, American Academy of the Political and Social Sciences, Philadelphia.

Rescher, N.: 1983, *Risk: A Philosophical Introduction*, University Press of America, Washington, D.C.

Ricci, Paolo F. and Molton, L.: 1981, 'Risk and Benefit in Environmental Law', *Science* **214**, 1096.

Rich, Robert F.: 1979, 'Systems of Analysis, Technology Assessment, and Bureaucratic Power', *American Behavioral Scientist*, January–February, 407.

Richmond, C. R., Walsh, P. I., and Copenhaver, E. P. (eds.): 1981, *Health Risk Analysis*, Franklin Institute Press, Philadelphia, Pennsylvania.

Rowe, William D.: 1977, *An Anatomy of Risk*, John Wiley & Sons, New York.

Rowe, W. B.: 1977, 'Government Regulation of Societal Risk', *The George Washington Law Review* **45**, 952.

Sassone, P. G. and Schaffer, W. A.: 1978, *Cost–Benefit Analysis: A Handbook*, Academic Press, New York.

Schneider, S. H.: 1979, 'Comparative Risk Assessment of Energy Systems', *Energy* **4**, 919–931.

Schwing, R. and Albers, W. A. (eds.): 1980, *Societal Risk Assessment*, Plenum, New York.

Shrader-Frechette, K.: 1980, 'Adams, Inhaber, and Risk–Benefit Analysis', *Research in Philosophy and Technology* **3**, 343–365.

Shrader-Frechette, K.: 1983, *Four Methodological Assumptions in Risk–Cost–Benefit Analysis*, National Technical Information Service, Springfield, Virginia.

Shrader-Frechette, K.: 1983, *Nuclear Power and Public Policy*, 2nd ed., Reidel, Boston.

Shrader-Frechette, K.: 1984, *Science Policy, Ethics, and Economic Methodology*, Reidel, Boston.

Slovic, Paul, Fischhoff, B., and Lichtenstein, S.: 1981, *Facts and Fears: Understanding Perceived Risk*, Briefing presented to General Electric, June 4, 1981.

Smith, K. R., Weyant, J., and Holdren, J. P.: 1975, 'Evaluation of Conventional Power Plants', Energy and Resources Program, Report ERG 75-5, University of California, Berkeley, California.

Soderstrom, Edward J.: 1981, *Social Impact Assessment: Experimental Methods and Approaches*, Praeger, New York.

Solomon, K. A. and Abraham, S. C.: 1980, 'The Index of Harm: A Useful Measure for Comparing Occupational Risk Across Industries', *Health Physics* **38**, 375–391.

Spangler, Miller B.: 1981, 'The Role of Syndrome Management and the Future of Nuclear Energy', *Risk Analysis* **1**, 179–188.

Starr, C.: 1979, *Current Issues in Energy*, Pergamon, New York.

Starr, C.: 1969, 'Social Benefit Versus Technological Risk: What Is Our Society Willing To Pay for Safety?', *Science* **165**, 1232–1238.

Starr, C., *et al.*, (1977), *Energy and the Environment: A Risk–Benefit Analysis*, Pergamon Press, New York.

Starr, C., *et al*: 1976, 'Philosophical Basis for Risk Analysis', *Annual Review of Energy* **1**, 629–662.

Starr, C. and Ritterbush, P.: 1979, *Science, Technology and the Human Prospect*, Pergamon, New York.

Starr, C. and Whipple, C.: 1980, 'Risks of Risk Decisions', *Science*, **208**, 1114–1119.

Tester, Frank J. and Mykes, W. (eds.): 1981, *Social Impact Assessment: Theory, Method, and Practice*, Detselig, Calgary, AB.

Thurow, Lester C.: 1980, *The Zero Sum Society: Distribution and the Possibilities for Economic Change*, Basic Books, New York.

Torgerson, Douglas: 1980, *Industrialization and Assessment: Social Impact Assessment as a Social Phenomenon*, York University Publications in Northern Studies, Toronto.

Ulvila, J. W. and Brown, R. Y.: 1982, 'Decision Analysis Comes of Age', *Harvard Business Review*, September–October, 130–141.

U.S. Nuclear Regulatory Commission, Atomic Energy Commission: 1975, *Reactor Safety Study: An Assessment of Accidental Risks in U.S. Commercial Nuclear Power Plants* (WASH-1400, NUREG 75/014), Washington, D.C.

Viscusi, W.: 1983, *Risk by Choice*, Harvard University Press, Cambridge.

Waiten, Cathy M.: 1981, *A Guide to Social Impact Assessment*. Report prepared for the Research Branch, Corporate Policy. Indian and Northern Affairs, Ottawa, Canada.

Ward, D. V.: 1978, *Biological Environmental Impact Studies: Theory and Methods*, Academic Press, New York.

Weaver, S.: 1979, 'Inhaber and the Limits of Cost–Benefit Analysis', *Regulation*, July/August, 24.

Whipple, C.: 1981, 'Energy Production Risks: What Perspective Should We Take?' *Risk Analysis* **1**, 29.

Whipple, C., Ricci, P., and Sagan, L. (eds.): 1983, *Technological Risk Assessment*, Sijthoff and Nordhoff, Winchester, Massachusetts.

White, G. and Haas, J. E.: 1975, *Assessment of Research on Natural Hazards*, MIT Press, Cambridge, Massachusetts.

White, Irvin L.: 1979, 'An Interdisciplinary Approach to Applied Policy Analysis', *Technological Forecasting and Social Change* **15**, 103–104.

Whyte, A. V. and Burton, I. (eds.): 1980, *Environmental Risk Assessment*, John Wiley & Sons, Chichester.

Wildavsky, Aaron: 1979, 'No Risk Is the Highest Risk of All', *American Scientist* **67**, 33–34.

Williams, C. A., Head, G. I., and Glendenning, G. W.: 1978, *Principles of Risk Management and Insurance*, American Institute for Property and Liability Underwriters, Malvern, Pennsylvania.

Wilson, R.: 1979, 'Analyzing the Daily Risks of Life', *Technology Review* **81**, 40–46.

Wodicka, Virgil O.: 1980, 'Discussion of the Regulation of Carcinogens', *Risk Analysis* **1**, 61.

Wolsko, R., Whitfield, R., Samsa, M., Habegger, L. S., Levine, E., and Tanzman, E.: 1981, *An Assessment of the Satellite Power System and Six Alternative Technologies*. Prepared for the U.S. Department of Energy and the National Aeronautics and Space Administration by Argonne National Laboratory. Report DOE/ER 0099.

Wynne, Brian: 1983, 'Redefining the Issues of Risk and Public Acceptance', *Futures* **15**, 13–32.

Yellin, J.: 1977, 'Judicial Review and Nuclear Power: Assessing the Risks of Environmental Catastrophe', *George Washington Law Review* **45**, 1977.

Zeckhauser, R., *et al.* (eds.): 1975, *Benefit–Cost and Policy Analysis*, Aldine Publishing Co., Chicago.

Zeckhauser, R.: 1975, 'Procedures for Valuing Lives', *Public Policy* **239**, 419–464.

INDEX OF NAMES

INDEX OF SUBJECTS